Critical Postmodernism in Human Movement,
Physical Education, and Sport

D0840190

SUNY Series on Sport, Culture, and Social Relations
Cheryl L. Cole and Michael A. Messner, Editors

Critical Postmodernism in Human Movement, Physical Education, and Sport

EDITED BY

Juan-Miguel Fernández-Balboa

State University of New York Press

Published by
State University of New York Press, Albany

©1997 State University of New York

For information, address State University of New York
Press, State University Plaza, Albany, N.Y., 12246

Production by Diane Ganeles
Marketing by Dana Yanulavich

Library of Congress Cataloging-in-Publication Data

Critical postmodernism in human movement, physical education, and
    sport / edited by Juan-Miguel Fernández-Balboa.
        p.   cm. — (SUNY series on sport, culture, and social
    relations)
        Includes bibliographical references (p.   ) and index.
        ISBN 0-7914-3515-6 (hc : alk. paper). — ISBN 0-7914-3516-4 (pb :
    alk. paper)
        1. Physical education and training—Social aspects.   2. Sports—
    Social aspects.   3. Human locomotion—Social aspects.   4. Critical
    pedagogy.   5. Postmodernism and education.   I. Fernández-Balboa,
    Juan-Miguel.   II. Series.
    GV342.27.C75   1997
    306.4′83—dc21   96-52319
                                                                                    CIP

10 9 8 7 6 5 4 3 2 1

Contents

PART I

The Human Movement Profession
in the Postmodern Era:
Critical Analyses

CHAPTER 1

Introduction:
The Human Movement Profession—
From Modernism to Postmodernism

Juan-Miguel Fernández-Balboa

Introduction

For the past three centuries, *modernism* has provided the dominant versions of political, economic, and social practice of the Western
civilization. At its center is the image of a coherent, rational "man"
[*sic*] who, through positivistic science and technology, has sought to
control Nature and constitute a totalizing and universal Truth. The
roots of modernism are found in the intellectual movement known as
the Enlightenment (Harpham 1994).

> [The Enlightenment's] major intellectual impulses were critical, ana-
> lytic, scientific, mechanistic, and anti-metaphysical. Epistemologi-
> cally . . . the deductive reasoning characteristic of the seventeenth-
> century metaphysics . . . was replaced by empiricist deduction.
> Politically, the Enlightenment is usually associated with social con-
> tract theory, individualism, natural right theory, and the pursuit of
> self-interest, rather than the search for community or the justifica-
> tion of hierarchy. (Jay 1984, p. 30)

Under the Enlightenment's philosophy the world has been dichoto-
mized (e.g., the West and the East, humans and nature, science and
metaphysics, mind and body). Moreover, within these pairs, there
exists a hierarchy. One of the elements of each pair is considered
superior to, and with automatic rights over, the other. Take, for instance, the duality West/East. Under a modernistic view based on
Enlightenment philosophy, the West is considered to be civilized,

sophisticated, advanced; the East, is seen as wild, exotic, primitive. Hence, the West is given the right to conquer, to imperialize; the East is supposed to welcome, to yield. To all this, and in relation to all other sociopolitical structures, we must add one more piece to the modernistic puzzle: a system of market economics resulting in "an increasing monolithic world in which nearly everything . . . is subsumed under a worldly universal principle—the monolith is Capitalism and the principle is profit" (Brosio 1993, p. 473).

All this has been justified under the premise (and the promise) of unrelenting progress. Progress was to be achieved through two imperatives of the modern society: efficiency in organization and humanitarism. On the one hand, aided by science and technology and at the expense of nature, "man" was to produce as much as possible; on the other hand, people were to recognize each other as equal in worth. These two modern imperatives, however, created an irreconcilable paradox: Efficiency of production requires a hierarchical organization, yet this type of organization is not egalitarian (Aron 1968).

There is another modern paradox: Success and happiness are left up to the individual, and yet, because of the mighty influence of powerful modern institutions such as schooling, the work place, and the mass media (Apple 1990; Parenti 1995), there is little sense of individual identity. Consequently, individual action has been rendered practically dormant. This situation has forced people into a twofold trap: (*a*) an uncritical faith and blind adherence to authority (be it political, religious, scientific, etc.) and (*b*) an almost total reliance on an economic system that artificially dictates consumer needs while producing and providing sources of personal identity and social connectedness (Fromm 1969). Under these circumstances, it is little wonder that individuals do not engage in political and civic-minded actions.

In the end, the modern order has been beneficial only for those at the top of the hierarchy (i.e., politicians and corporate tycoons who have taken advantage of their privileged position and grown extremely rich and ever more powerful). The rest of the people, in contrast, have sadly sunk into lower and lower levels of hopelessness in a world where crime, war, hunger, poverty, pandemics, environmental destruction, and so forth, have been the norm. As a result, "Many groups [of people] now share a sense of deep alienation, despair, uncertainty, loss of a sense of grounding even if it is not informed by shared circumstance" (bell hooks 1990, p. 27). These are clear signs of an acute global crisis, a crisis brought about by modern thought and action.

Nevertheless, the status of things is never quo. The dominant groups can be challenged and their power eroded. In this regard, and as a result of today's disenchantment with the "modern project," many people have begun to wish for a new era—an era in which equality, dignity, and hope, as well as a strong sense of identity and community are more the norm than the exception. Indeed, there seems to be evidence that we might have already started such a transition—a transition into the so-called postmodern era (e.g., Agger 1992a; Lyotard 1992; Seidman 1994).

Postmodern Delineations and Critical Theory

A general consensus has not yet been reached about what constitutes postmodernism. Some theorists believe that postmodernism is a radical break from modern habits, relations, and social practices that rejects traditional narratives and any other form of totalizing thought (Lyotard 1994). Yet, others hold the opinion that "postmodernism cannot be a simple rejection of modernity; rather, it involves a different modulation of its themes and categories" (Laclau 1988, p. 65). The debate still continues.

Most theorists, however, recognize that postmodernism is more than a mere chronological transition (say, from industrialism to post-industrialism). The shift from the modern to the postmodern encompasses cultural critique, political activism, praxis (Freire 1970). From a postmodern perspective, culture is no longer perceived as a process toward progress nor as a linear historical trajectory of humans toward some predetermined end. According to Agger (1992b), "postmodernism is a theory of cultural, intellectual, and societal discontinuity that rejects the linearism of the Enlightenment notions of progress" (p. 93). As such, postmodern culture seems to encompass various ways of social organization in which new forms of language, cultural assumptions and meanings, social movements, and power relations can emerge (Butler 1994).

Postmodern theorists have embraced two main ideas: That of subjectivity (i.e., the personal is political) and that of knowledge as power (Foucault 1980). In this vein, political and social activism and contestation are understood within a framework of civic competence and the admission of different moral affirmations. As such, identities, meanings, and relations are not seen as fixed and constant but, on the contrary, as "finite, locally determined language games, each with specific pragmatic criteria of appropriateness or valence"

(Lyotard 1984, p. xxiv). Similarly, knowledge and "reality" are perceived as governed by linguistic codes organized and categorized in particular ways that are beneficial to some and detrimental to others (Cherrylholmes 1988; Vygotsky 1978). "One of the most basic themes of postmodern debate revolves around reality, or lack of reality or multiplicity or realities . . . in which previous modes of social analysis and political practice are called into question" (Lyon 1994, p. 7). Moreover, in order to be consistent with its own principles of critique and deconstruction (Derrida 1976), critical postmodern theory must have an internal method of self-interrogation in which assumptions of oppression and freedom are open to reformulation. Otherwise, postmodern theory would be no different from the modern ideology that it seeks to scrutinize.

Postmodernism, Critical Theory, and the Human Movement Profession

Postmodern theory and thought did not emerge in a vacuum. The basis for postmodern sociopolitico-cultural delineations can be found in critical theory. Critical theory, originally formulated in the 1930s by the founders of the Frankfurt School (i.e., Marcuse, Adorno, Horkheimer) and later extended by others (e.g., Habermas, Foucault, Fromm), has been variously characterized as a radical theory of cultural criticism that analyzes a number of social problems emerging from the Enlightenment. In order to develop workable models for a future better society, critical theorists examine "things, institutions, practices, and discourses [while seeking] an autonomous, non-centralised kind of theoretical production, one . . . whose validity is not dependent on the approval of the established régimes of thought" (Foucault 1994, p. 40). Thus, more akin to moral philosophy than to predictive science, critical theory offers an alternative approach to understanding the ideologies behind modern social, cultural, economic, and political formations and assumptions (Ingram and Simon-Ingram 1991).

In this regard, those who believe that our society is undergoing a transition from modern to postmodern principles, will agree (a) that living, teaching, and learning in this new society will require new concepts, attitudes, and actions; (b) that institutions will need new schemes; and (c) that the means of knowledge production and the criteria for validating such knowledge will have to be reconceptualized in the sense that one no longer will be able to claim the exclusive supremacy of one single canon or truth. All this has important implications for the professions.

In this book, the authors examine the transition from the modern to the postmodern as it affects the human movement profession, physical education, and sport. For the sake of clarity and practicality, we use the phrase "human movement profession" as an umbrella term that encompasses many professional groups (e.g., athletes, educators, coaches, sport administrators, recreation leaders, researchers) dedicated to the practice, pedagogy, and study of human movement and its related activities.

Our analysis of the profession of human movement is framed within critical postmodern theory. As the readers will see, the authors of this book display different versions and interpretations regarding critical postmodern theory. Yet, we believe that dissension and polivocality are legitimate—after all, they are within the spirit of postmodern critique. On the other hand, notwithstanding our different views, we have a common intention. We set out to delineate ideological and political markers to enable readers to see the profession, not in a vacuum, but as a political movement of sorts; a diverse collection of communities, and a forum for acceptance, equality, and freedom.

With this intention, we analyze the origins and the processes of construction, regulation, distribution, and legitimating of the master narratives of the modern human movement discourse. We attempt to uncover how, by muddling our vision and creating false images of reality, these master narratives have embodied, and still do, particular epistemological and political views which benefit a few at the expense of many. We argue that these master narratives have been "largely drawn from cultural scripts written by white males whose work is often privileged as a model of high culture informed by an elite sensibility that sets it off from what is often dismissed as popular or mass culture" (Aronowitz and Giroux 1991, p. 58). These master narratives, we also argue, have been based on a world view that exults "continual progress of the sciences and of techniques, the rational division of industrial work, and the intensification of human labor and or human domination over nature" (Baudrillard 1987, p. 65–66).

Our quest for new professional grounds and possibilities forces us to ask critical questions with regards to our traditional roles and purposes in the field of human movement. Namely, What are the conflicts and crises facing our profession? How can our professional relations and meanings be constructed within a new realm of justice, freedom, and equity? What moral and liberating principles and practices are needed to give our profession an identity of a collective, more encompassing self? How can we reinterpret, create, and utilize our professional knowledge? In what ways can we teach this new knowledge so that it becomes both emancipatory and liberating? What should the basis for our future professional actions be?

No doubt, the answers to these questions depend on one's historical, intellectual, and cultural position. In exploring new cultural, political, pedagogical, and scientific professional paths, a plurality of narratives and a polyphony of voices must be taken into account. Furthermore, we are conscious that our exploration must be exercised with caution. We recognize that all of us are affected by various ideological distortions and discourses. This is why this book has been divided into two parts.

In part I of this book (i.e., The Human Movement Profession in the Postmodern Era: Critical Analyses), the authors want to raise awareness about the limitations of modernism and the possibilities of postmodernism with regard to human movement, physical education, and sport. One of the central arguments here is that the profession of human movement is not isolated from broader social, political, and cultural contexts and histories, but, instead, it is very much related to, and affected by them. Historically this profession was born and has evolved within the frame of modernism and, therefore, it is still being influenced by the same powers, confined by the same institutions, affected by the same problems, and determined by the same ideological principles and values that have molded modern life. In turn, the human movement profession has shaped people's meanings and practices regarding physical activity. That is, the modern approach to human movement has influenced the ways people view their bodies, exercise, and create and use movement-related knowledge. We believe that these ways have generally suited the interests of dominant social, economic, and political groups while silencing and discriminating against those with less power. Therefore, in part I, we undertake to critique these conventional ways and to offer alternative ones.

Part II of this book (i.e., Critiques of the Critical Postmodern Analyses of the Human Movement Profession) contains commentaries on, and critiques about, the analyses presented in part I. The authors in part II act as checkers and balancers to the theories and actions proposed by the authors in the previous section. However, they do not limit their writing to mere commentaries and critiques. They, too, reveal their perspectives and bring to the forefront their ideological values and professional practices. In this regard, the ideas presented in part II encompass a larger number of experiences and beliefs that, rather than to polarize, serve to expand our options for future praxis in human movement, physical education, and sport.

In one word, this book is an honest attempt to push our profession forward in a unifying manner, not in an antagonistic one. Our objective is not simply to unveil alternative perspectives and expand professional boundaries, per se, for doing so would place us in a position of authority which we do not wish to assume. On the contrary, we are aware that what we present here are just theories in the making that, notwithstanding, go beyond criticism and argumentation by offering what we believe to be valuable and viable ways for action and reflection. In so doing, we move into the sphere of praxis and adopt strong political stances that reflect our diverse struggles on different fronts within the profession. In this sense, the book reflects our strong commitment to delineating new borders, offering new hope, and creating a better future for us all.

CHAPTER 2

Sociocultural Aspects of Human Movement:
The Heritage of Modernism,
the Need for a Postmodernism

George H. Sage

Introduction

A book devoted to the human movement profession and the turn
from modernism to postmodernism, as this one is, requires a review
of the past and present sociocultural conditions. We must know back-
ward to think forward. Social change to a postmodern society must
be grounded upon an understanding of the traditions of modernity—
not just for knowledge sake, but rather to remedy the oppression,
injustices, and undemocratic traditions which modernism has perpetuated in the name of reason and social progress.

Two aims underlie my analysis of the sociocultural aspects of
human movement, in particular in the areas of physical education
and sport. First, because physical education and sport are consti-
tuents of the larger society, I examine the broad sociohistorical, political, and economic dimensions which characterize modern society.
I do so from a critical theoretical perspective which begins at a common point: the realities of the enormous social inequalities in power,
wealth, prestige, and opportunity, not only in the realms of gender,
race, class, and so forth, but also within the nondemocratic institutional traditions of modernity.[1] This approach is quite different from
mainstream analyses of modern societal conditions, which emphasize harmony, consensus, and "functional" social interrelationships
in modern society.

Second, I want to elucidate the symbolic and cultural meanings
of physical education and sport. Here, I will analyze the connections

11

between physical education, sporting practices, and society. At the same time, I will focus on the production and maintenance of ideologies of dominance and subordination typical of modernity and the role of human movement in the dynamics of social reproduction and change.

Before proceeding, two cautionary notes are necessary. First, I cannot hope to capture the full extent of the historical diversity of the people that will be discussed in this chapter. My descriptions and accounts will have a distanced focus. Second, given the historical and conceptual ties between physical education and sport, I will examine both concurrently.

The Crisis of Modernity

Numerous social and cultural analysts now claim that the culture of the Enlightenment, which has formed the modern political, economic, intellectual, and social foundations of the Western societies during the past centuries, is in crisis. Assumptions about "the unity of humanity, the individual as the creative force of society and history, the superiority of the West, the idea of science as Truth, and the belief in social progress" (Seidman 1994, p. 1), all characteristics of the culture of the Enlightenment, are being challenged by broad social, intellectual, and cultural alternatives (Baudrillard 1988; Bauman 1993; Jameson 1990a; Lash and Urry 1987; Lyotard 1984, 1993). The postmodern concept is applied to this turn away from modernity and Enlightenment assumptions. As such, postmodernism may be viewed as the possible next stage of social history because it eschews the failed ideologies and practices of modernism and aims to bring about a more democratic and socially just world.

In chapter 1, Fernández-Balboa emphasizes that the profession of human movement is not isolated from the broad political, economic, and cultural conditions in which it is embedded. Therefore, attempting to understand postmodern perspectives and prospects necessitates grounding them in the sociocultural traditions of modernity and the culture of the Enlightenment. After all, postmodernist cultures do arise from the traditions of modernity. "The chief signs of modernity have not disappeared. . . . Modernity has not exhausted itself; it may be in crisis but it continues to shape the contours of our lives" (Seidman 1994, p. 1).

Mapping the Historical and Current Social Oppressive Features in Modern Society

Widespread inequality, oppression, and social injustice in terms of class, gender, and race, are fundamental and overriding characteristics of the so-called modern era. Hence, an analysis of the sociocultural aspects of physical education and sport must historically situate modernist traditions and institutions while denouncing the social inequalities based on class, sex, race, age, sexual orientation, physical ability, and so forth. Fundamental to these inequalities are the power relations that have been socially and historically constituted throughout modernism.

In the following subsections, I examine class, gender, and racial oppression and injustice within the context of two major social institutions which exist in every modern nation and across time: the polity and the economy. Ideally, the former is concerned with the promotion of the social order and general welfare, while the latter attends to the production and distribution of goods and services.[2]

Political Inequality and Oppression

Modern democratic politics began with what some have called the "two great political revolutions" of the eighteenth century: the French and American revolutions. National democracies found throughout the world today are rooted in these two revolutions. But it is a mistake to believe that their immediate outcome, or even their ultimate consequences, was an egalitarian and democratic social order. The framers of the Constitution of the United States, rich and powerful white men, were not champions of the common people; instead, they were committed to retaining their wealth and privilege by keeping common people, whom they considered ignorant and threatening, subordinated (Fresia 1988). For example, Alexander Hamilton called the people "a great beast." Moreover, as originally written, the Constitution barred most people (i.e., women, slaves, Native Americans, and the poor) from voting. Fresia (1988) asserts that "at the very heart of [American] political institutions, at the very core of our way of doing politics is fear and distrust of the political activity of common people" (p. 9).

Analyses of the political heritage of other democracies around the world reveal similar, even more profound in some cases, antidemocratic traditions. More important, many democratic states have at one time or another been ruled by one-party governments (e.g., Mexico, the Philippines), or worse, by military dictatorships. Although individual struggles for democratic institutions and opposition and resistance to anti-democratic conditions have been a constant fact of life for people throughout the modern world, freedom and democracy have proven to be elusive and fragile.

CLASS INEQUALITIES. Some postmodernists have argued that class analyses are outdated and that the designation of a "working class" is less valid in a postmodern era than it was during the industrial era (Lash and Urry 1987). Other postmodernists, however, defend such analyses on the basis that the overwhelming majority of adults still are wage workers and, therefore, beholden to employers. They argue that because exploitation of working people and class struggle continue to exist, the concept of class must be examined within postmodern analyses (Bauman 1987; Harvey 1989; Jameson 1988; Soja 1987). Harvey (1989) explains that there are many continuities between modernity and postmodernity, and although there has been a "change in the surface of capitalism . . . the underlying logic of capitalist accumulation and its crisis tendencies remain the same" (p. 189). My own preference leans towards the latter argument.

From the beginning of the modern era in the latter eighteenth century, the dominant political groups in Western countries have been white, male, upper class, and Christian. Outside the West, wealthy males have reigned supreme, as well. These dominant groups own the land, capital, and technology and employ most of the labor force. Moreover, these groups have articulated their version of "social reality" to the majority of the population, thus allowing this version to represent the "national interest." In every modern society, these dominant groups have acquired and influenced political power to protect and foster their own interests.

GENDER INEQUALITIES. The dominant gender ideology throughout the modern era has been patriarchy, which has portrayed women as inferior to and dependent on men. Women's primary role prescriptions have been as childbearers, childrearers, homemakers, and sex objects. Hence, patriarchy has been the fundamental political factor which has shaped women's lives in all social facets.

Patriarchal traditions of modernity have been perpetuated in the United States with respect to the political status of women. The framers of the U. S. Constitution did not extend to women the right to vote. It was not until 144 years later that women gained this right (19th Amendment to the Constitution). Many other modern states have also constructed barriers of various kinds to discourage or prevent women from exercising their democratic rights (e.g., Hause 1984; Kolinsky 1989). Thus, discriminatory cultural definitions of gender have been embedded in the political traditions of modern Western societies.

Still, acquiring the right to vote has neither guaranteed women access to political institutions in Western nation states nor has it provided them with the kind of social support that helps women in their domestic or occupational roles. In all the "democracies" in the West, political positions at all levels of government have remained dominated by white, upper-class males. For example, in the U.S. only 11 percent of the House of Representatives are women, and only eight U.S. senators are female. Also, less than 8 percent of all federal and state judges are female, fewer than 6 percent of all partners in law firms are female, and less than 1 percent of top corporate managers are female. Similar conditions exist in other so-called democratic states throughout the modern world.

The marginal status of women in the political sector reaffirms the necessity for the struggle that feminists and others are waging for women's rightful place in the corridors of power. Until women can equally make, interpret, and enforce the laws of their nation, they will not have their interests adequately represented.

RACIAL INEQUALITIES. People of color have had to cope with discrimination based on race for centuries. As Morrow and Torres (1994) note: "Racism goes back to its precapitalist roots and has been associated with many different types of social formations, though closely associated with slavery and colonial expansion in its modern form" (p. 55). Racism has been rampant in the modern world.

One of the most abhorrent examples of racism during the modern era is embedded in America's history. The Declaration of Independence and the U.S. Constitution condoned racial subordination of Blacks. In spite of their enlightened stance toward freedom and liberty, the framers of these documents did not see them as applying to blacks. In fact, slavery was sanctioned and Blacks were denied the right to vote. Although slavery in the U.S. was abolished in 1863, Jim Crow laws—segregation laws which legalized white domination—left racism essentially intact. The "separate-but-equal" doctrine promulgated by the U.S. Supreme Court became a new regime of discourse[3] that constructed the meaning of race and became an efficient instrument of domination and subordination. Only in the past twenty years have the civil rights of black citizens been protected by law. But despite these laws and improved conditions in some private and public spheres, the subordination of Blacks is still a basic attribute of American society (Carnoy 1994).

Of course, every modern nation has its own specific forms of racial oppression sanctioned in various ways by the political apparatus. Many countries have laws restricting the personal freedoms and other human rights of diverse ethnic and racial groups. Some nations provide these people with little protection from abuse and attack; in most governments, social services are either denied, or not provided to these groups to the same extent that they are to other citizens.

Postmodernism and Politics

As will become evident in the chapters of this book, postmodern politics "speaks to the needs of the marginalized Other—to women, to people of color, to gays and lesbians, to the working class, to the poor and homeless" (Haber 1994, p. 3). From a postmodern perspective, the extent to which a political regime is considered viable is the extent to which it recognizes and pays attention to difference—to those who have traditionally been mistreated, silenced, oppressed, and segregated.

Economic Inequality and Oppression

Virtually simultaneous with the two great political revolutions in France and Colonial America, a third, but economic, revolution began: the Industrial Revolution. Starting in England in the late eighteenth century, it spread in the nineteenth century throughout Europe and North America, becoming the economic foundation upon which the modern world has been constructed. In addition to the technological innovations ushered in by the Industrial Revolution, broad social and cultural transformations were also fostered. During the early decades of the nineteenth century, most of the working population in Western countries were small independent farmers, skilled artisans and craftsmen, small retail shopkeepers and merchants, and shippers and traders. Daily economic activities centered around the seasons and harvests and bonds of social interdependence. This changed dramatically with the Industrial Revolution and its economic companion, capitalism. The capitalist mode of production grew in prominence, and a factory system developed in England and then spread to other Western nations. This promoted the growth of cities on a scale unheard of in history, agrarian societies and cultures gave way to urbanized ones, and new economic inequalities emerged.

CLASS INEQUALITIES. With industrial expansion, the labor power to run the factories and carry out the semiskilled and manual tasks to keep industries growing was supplied by wage workers. The proportion of the labor force engaged in wage labor in the U. S., for example, steadily increased from approximately 20 percent in 1840 to 72 percent in 1910 (Gordon, Edwards, and Reich 1982). Presently some 90 percent of working Americans are wage earners or salaried employees, increasingly in information and service occupations. This pattern from preindustrial to industrial has been repeated throughout the modern world.

In the capitalist labor structure, the average worker is rather powerless. Workers without material goods or the material means of production must depend on their labor power, their ability to work. This system of relations results in workers without property rights to the products of their labor (Lebowitz 1988). Inherent in capitalist ownership of property is legal control over its use. Hence, as Steinberg (1982) says, the employee faces "inequality in the labor market armed only with a set of legal rights more appropriate to a preindustrial society in which work was carried out by independent entrepreneurs and skilled craftsmen" (p. 4). Carnoy and Shearer (1980) concur, arguing that the "the work place [is] governed by the laws of private property, not the Bill of Rights" (p. 12). Democratic workplaces, with workers participating in the decision process and evaluating their work in terms of its social significance or its moral effects, have always been viewed as impractical and quixotic, from the perspective of management.

Workers' share of the value they create through their labor power has never been equitable. Today in the U.S., 1 percent of the population—2.5 million people—earns more income than the poorest 40 percent—100 million people (Bohmer 1992; Miller 1995; Perelman 1993); meanwhile, thirty-five million Americans live below the poverty line and homelessness is on the rise (Timmer, Eitzen, and Talley 1994). While the family income of the wealthiest Americans increased dramatically over the past fifteen years, the number of full-time workers in the U.S., who are impoverished, has increased by 50 percent; 18 percent of full-time workers now fall below the poverty line (Freeman and Katz 1994; Griffith 1993; Scheer 1994). The United States "has the highest incidence of poverty among the non-elderly and the widest distribution of poverty across all age and family groups"; it is also "the only Western democracy that has failed to give a significant portion of its poor a measure of income security" (McFate 1991, p. 2). Rather than encouraging a redistribution of wealth in the U.S., the rich get richer and the poor get poorer (Braun 1991; Dembo and Morehouse 1994; Wolff 1995).

Economic trends strongly suggest that conditions are worsening for working people throughout the world. During the past three decades, national economies have been eclipsed by the globalized economy. Taking advantage of postmodern technology, First World corporations have moved to Third World countries where it is easier to exploit workers. While the public discourse throughout the world is on the economic benefits of the capitalist globalized social order, in reality, workers throughout the world face enormous inequality. In 1993, the unemployment rate of the twelve countries of the European Community was a distressing 10.3 percent (Barnet and Cavanagh 1994). According to the International Labor Organization some 30 percent of the world's active labor force (820 million people) are currently unemployed or underemployed, and almost one quarter of the world's population, about 1.2 billion people, live in absolute poverty (Epstein, Graham, and Nembhard 1993; Peterson 1994; World Bank 1990; Worldwatch Institute 1990). Furthermore, in the past thirty years the per capita income gap between the developed and underdeveloped countries has actually widened (Arrighi 1991; Worldwatch Institute 1990).

GENDER INEQUALITIES. One reason that capitalist economies look the way they do is because of their patriarchal relations (Wyss and Balakrishnan 1993). Ironically, until quite recently, women were written out of economic accounts of labor history. This absence conveyed the impression that women were not participants in the workplace when, in fact, they have been ubiquitous in the industrial labor force for over two hundred years. Even more important is that, despite the invaluable contribution that it makes beyond the household to the larger capitalist economic system, the unpaid domestic labor of women, in general, has never been recognized (Hochschild and Machung 1989; Palmer 1989; South and Spitze 1994).

Today, women in developed countries make up 40 to 60 percent of the labor force; they constitute the bulk of the labor force in Third World countries throughout the world. In the *maquilas* of Mexico, for instance, "young women of rural origin make up 70 percent of the labor force of half a million workers" (Rodríguez 1993, p. 296; also see Fernández-Kelly 1983). Likewise, in Asia, females between sixteen and twenty-five years of age make up over 65 percent of the export processing work force. Many are unable to continue working beyond age twenty-five because of work-related health problems brought on by long hours and unhealthy working conditions (Kamel 1990; LaBotz 1992; Ward 1990; Wolf 1992).

Also as a norm in the capitalist system, women everywhere are paid lower wages than men because they are perceived as being docile and obedient—less likely to protest against low pay and poor working conditions. Ironically, it is their low wages and long hours which makes capitalist firms commercially successful. The reality is that, as production of consumer goods have become increasingly globalized, the exploitation of women has become globalized, too (Barnet and Cavanagh 1994; Wyss and Balakrishnan 1993).

At the same time that women struggle in menial and low-paying occupations; many jobs, even entire occupations, are considered to be inappropriate for them by the male-framed division of labor. Also, sexual harassment is an every-day experience for many women in the workplace. It comes from bosses, supervisors, and coworkers who use their position to coerce, intimidate, and threaten female employees (Tomaskovic-Devey 1993).

RACIAL INEQUALITIES. An unequal relationship endemic to the capitalist mode of production involves Blacks and ethnic groups other than white. Racism has been intrinsic in wage labor throughout history. In 1619, Blacks were first brought to Colonial America to serve as an unpaid source of labor for both agricultural and commercial interests. Slave owners of the preindustrial agricultural South, together with Northern trading and shipping firms, created a racist social structure with Blacks at the bottom. After the Civil War, Blacks were unequally indentured into capitalist wage labor.

Blacks have always been discriminated against in terms of job opportunities, job advancement, wages, and working conditions. In the U.S., during the past twenty years, the economic gap between whites and blacks has actually widened. Presently, among the nation's Blacks, the poverty rate is 33 percent, three times that of Whites. Black family income is only 56 percent of Whites (and *declining* in the past ten years). Blacks are twice as likely to be unemployed than Whites, and when employed are mostly in jobs where pay, power, and prestige are low. Only a smattering of Black managers has moved beyond middle levels of authority and control. Less than 3 percent of American physicians, dentists, and pharmacists are Black. The widespread underemployment and unemployment that African-American men experience creates a cultural scenario in their communities that reinforces the general public's stereotype of the lazy, shiftless black male (Tomaskovic-Devey 1993).

Other countries of the world have their heritage of racial bigotry and oppression, as well. South Africa had an institutional system of apartheid for decades, and Black workers labored under the most inhumane conditions imaginable. Western European countries have relied on "guest workers" to take on the lower paying jobs. In the same way, Japan has employed Asian immigrants since the end of World War II. These workers have been employed to do the most menial, dirty, and low-paying jobs, while facing various forms of discrimination and exploitation in the workplace (Herbert 1990; Shimada 1994).

Postmodernism and the Economy

Modernity has advanced under the banner of a free-market economic liberalism, promising economic prosperity, but it has delivered worldwide impoverishment and created an unconscionable gap between the "haves" and "have-nots." In this sense, modernity has not turned out to be a force of human liberation ; instead, it has become a force limiting human material conditions and obstructing political and economic progress. As an alternative to the modern system, in his book *Explorations at the Edge of Time*, Falk (1992) advances a model of a postmodern future consisting of numerous international social and cultural forces committed to human rights and a vision of human community based on the harmony of diverse cultures striving to end poverty, oppression, and injustices of all kinds. Falk's approach is both humanistic and cosmopolitan; it is also a celebration of plurality and economic rights that includes some kind of redistribution of resources and wealth whereby workers are assured decent living wages, a comfortable standard of living, health care, educational opportunities, and other basic social services.

According to Jameson (1990b), in postmodern society, "democracy must involve more than political consultation. There must be forms of economic democracy and popular control in other ways" (p. 31). Dagnino (1993) agrees, arguing that a goal of postmodernist social movements will be the creation of an alternative enlarged definition of democracy which includes the workplace and other human social practices, not just the State.

Modern Symbolic and Cultural Meanings
of Human Movement Practices

During the modern era, the domains of education and popular culture have been crucial areas for the reproduction of ideologies of dominance and subordination. Educational and popular culture symbolisms have created legitimating rituals and ceremonies and reinforced dominant cultural practices. Among these cultural practices are those related to human movement (e.g., physical education and sport). Because physical education and sport are forms of popular culture, any notion that they are separate from the cultural processes of society at large is naïve and misguided. In fact, physical education and sport in all of their forms have been fully integrated into the power structure and social relations of modern societies (Kirk, chapter 4).

One of the most fundamental functions of human movement practices in the modern era has been to promote initiatives and activities that help shape the economic, political, and cultural hegemonic structures of the dominant groups. Compelling evidence shows that the inequalities and injustices of power and privilege that have existed in the wider society have been consistently patterned and persistent over time in human movement practices. Human movement's symbolic importance, then, is rooted in its ability to promote and structure relations in accordance with the needs and interests of dominant groups. Put another way, the practices and values surrounding human movement continually reinforce the dominant interests (Hargreaves 1986; Kirk 1993; Sage 1990).

Modern Symbolic and Cultural Meanings:
Human Movement and Polity

One traditional purpose of the modern state involves fostering and preserving social harmony by generating a sense of loyalty, nationalism, and patriotism among its citizens. It does this by creating and maintaining social stability and support for the reigning groups. In this regard, the cultural activities related to human movement serve as a national vehicle for advancing the prevailing social order through the use of symbolic representations in rituals and ceremonies. The political apparatus conveys particular symbolic codes that reinforce the dominant culture and its own world view. Human movement (through both educational, leisure-like, and professional sporting events) constitutes a focal point for formal ritual and ceremony centering around governmental symbols. Important national and international events like the Super Bowl, World Cup, and Olympic Games are incorporated into a symbolic panoply of political rituals that serve to remind people of their national identity and "common" destiny. Such events help create and support the political order by conveying messages about norms, values, and dispositions that reinforce the ideological hegemony of dominant groups. Thus, there is formidable evidence that human movement activities are not neutral in their relations with the State; indeed, political ideologies are inscribed in the human movement culture.

Modern Symbolic and Cultural Meanings:
Human Movement and the Economy

In the West, human movement programs in physical education and school sports originated in the nineteenth century (see also Kirk, chapter 4, for an Australian version of this issue). The former became part of the national systems of education that many industrialized nations were building. At the same time, school sports were promoted by educators and industrial leaders who were concerned with "building character" in an expanding entrepreneurial environment (Mangan 1975, 1986; Miracle and Rees 1994). By the turn of the century, the expansion of formal human movement education and organized sport was seen as a means of integrating the growing industrial working class into an expanding capitalist order—an authoritarian, hierarchically organized, and rational order (O'Hanlon 1980; Spring 1974).

Under the capitalist system, goals were to be achieved as completely and cheaply as possible. Hence, rational organizations were designed to minimize the discretionary behavior of individuals by making organizational processes more routine and predictable. State-supported educational systems (as well as private schools), many of which included physical education and interschool sports, were organized around obedience to the authoritarian and hierarchical order and designed to inculcate discipline and hard-working habits—the same values and behaviors of the capitalist workplace (Mangan 1986; Miracle and Rees 1994).

Modern Symbolism and Cultural Meanings:
Human Movement and Social Class

As forms of cultural practice, human movement practices such as physical education and sport promote the social class inequality so endemic to capitalism (Sage 1990). Their role in reproducing the class structure is played largely through the socialization of participants and spectators into the legitimacy of the capitalist system. By teaching capitalist values and meanings and by legitimating the existing system of societal rewards and privileges, movement activities "prepare young men [and increasingly young women] to take for granted the norms of the capitalist workplace; and central among these is that every aspect of the process is necessarily geared to the 'natural' goal of increasing productivity" (Whitson 1986, p. 101). Young athletes and students, then, may be viewed as engaged in a form of "anticipatory productive labour" (Fine 1987; Ingham and Hardy 1984).

Furthermore, human movement practices symbolically reflect and promote the modernistic ideology through their glorification of meritocratic standards of hierarchy and success based on skill, the valorization of commercialism, and the presentation of a false view of social progress, which emphasizes records and performance standards. On this point, Gruneau (1982) states:

> . . . much of the organization and culture of modern sport seems to have been influenced by capitalist productive forces and relations. For example, "amateur" sports at their highest levels have almost become monuments to such new sciences as biomechanics, exercise physiology, and sport psychology where a market rationality is expressed in a mechanical quest for efficiency in human performance that is indentured to state and commercial sponsorship. Professional sports, meanwhile, have gone a great distance toward reducing the meaning of athletic contests to a simple dramatization of commodity relations. (p. 24)

At the same time, human movement practices serve to promote and sustain an hegemonic ideology regarding widespread social mobility in the larger social structure. One of the most deep-seated beliefs in modern capitalist states is that they are egalitarian, socially-mobile societies in which everyone can reach the top. Indeed, a widespread creed is that of "equality of opportunity." This belief is reinforced in the U.S. by the "rags to riches" stories, the theme of which is people who have been born into poverty can, by applying themselves to their work and saving their money, rise through sheer effort to a position of social, economic, and occupational importance. Sport stars are an example of this.

The "equal opportunity" ideology is then incorporated into the so-called work ethic orientation towards social mobility. The essence of this orientation is that individuals are responsible for their own fates, so those who want to get ahead can get ahead; all that is necessary is hard work and a determined willingness to strive for success. Here, it is assumed that those who do work hard will, in fact, become successful in material terms; while those who do not, will not succeed. Implicit here is that the latter have only themselves to blame for their lack of dedication.

Human movement is a strong contributor to this ideology. In fact, through hidden and overt curricula, human movement practices powerfully contribute to the ideology of a competitive and meritocratic social order (Fernández-Balboa 1993a). Because competitive human movement activities, by their very nature, bring status and rewards according to supreme performance, they provide convincing symbolic support for the myth that dedication, sacrifice, and hard work will bring upward social mobility. As one example, the few athletes who become professionals or high performance amateur athletes reinforce the public's belief that the social class system is more open to social mobility than it really is.

Modern Symbolism and Cultural Meanings: Human Movement and Gender

Historically, competitive movement activities have been a male preserve. Gender difference has been constructed through sporting activities and the human movement culture. Messner (1992) has argued that from its beginning, modern sport "was constructed as a homosocial world, with a male-dominant division of labor which excluded women. Indeed, sport came to symbolize the masculine structure of power over women" (p. 16). He has also pointed out that the hegemonic masculine conception of physically demanding competitive activities bonds men, symbolically, as distinct and superior to women. "It is this simultaneous expression of difference and bonding among men that defines the role that modern sport plays in the contemporary gender order" (p. 19).

Because the structure, values, and ideology of human movement are profoundly gendered, advantage and disadvantage, control and exploitation, and meaning and identity pattern the distinctions between male and female, the masculine and the feminine. Experiences in human movement activities constitute a gendering process whereby boys and men learn the dominant cultural expectations of what it means to be male (Clatterbaugh 1990; Messner and Sabo 1994; Miedzian 1991). At the same time, girls and women learn what it means to be female. As one example, the sex-segregated nature of physical education programs and many sporting activities "provide a context in which gendered identities and separate 'gendered cultures' develop and come to appear natural" (Messner 1992, p. 31). Similarly, physical education and school sports support the *status quo* by reproducing images of masculinity and femininity that sustain the same asymmetrical gender relations of the larger society (Birrell and Cole 1994; Costa and Guthrie 1994).

Through human movement activities, hegemonic masculinity has become part of our commonsense understanding of the world, legitimating not only inequities within the realm of human movement activities but also patriarchal relationships generally. In this regard, S. Cahn (1994) has argued:

> More important, these arrangements shape the contours of gender relations in the wider society, contributing to notions of "natural" male superiority, immutable sexual differences, and normative concepts of manhood and womanhood. (p. 223)

Modern Symbolism and Cultural Meanings: Human Movement and Race

It is the stereotyping of people of color that symbolically oppresses them, creating cultural meanings that "naturalize" differences between Whites and Blacks. Characterizations such as "lazy," "unintelligent," "undependable," "lacking courage in the face of adversity" (there are many others) are part of a regime of discourse that has been used in the modern society to portray people of color as inferior and, hence, undeserving of equitable treatment. These stereotypes have been employed to establish unequal racial boundaries in the context of human movement. Physical education and sport practices, for instance, reinforce and reproduce attitudes and beliefs about people of color.

In most countries of the world where Whites dominate, Blacks and other minorities have suffered multiple patterns of discrimination in the human movement culture. Although the specific patterns vary, there is a common pattern of first prohibiting Blacks and other ethnic minorities from participating in certain forms of human movement (e.g., sports), then permitting their involvement, but on a segregated basis, and finally integrating them into these practices, but limiting their participation in certain ways (e.g., limiting the scope of human movement activities or play to only certain positions in sports). In the latter case, Blacks and ethnic minorities have sometimes been "stacked"—been assigned to certain playing positions and denied access to others (Hallinan 1991; Lavoie 1989; Maguire 1988; Schneider and Eitzen 1986).

Although it is true that access to physical educational opportunities and to playing sports has expanded greatly for people of color in recent years, the symbols of racism still remain, and management and administrative positions in education and sports continue to elude the members of these racial groups. Because the higher administrative and leadership positions, with the greatest power, prestige, and material rewards, are more insulated from direct scrutiny, those who control the structural organization and "job" assignments and access to higher management and administrative levels, can subtly continue discriminatory practices. In this regard, Blacks and other "minorities" still have difficulty securing the powerful, higher-paying, and prestigious positions in sport.

Closing Remarks: Symbolism, Cultural Meanings, Movement, and Postmodernism

Human history is not a mono-linear unfolding towards a predetermined end; instead, it is an active constructing process. Human movement is intricately linked to the larger society in which it is embedded. The possibilities of the postmodern turn suggests that human movement professionals can play an integral role in the social transformation of the modernistic political, economic, and cultural inequalities and injustices of power and privilege. By making problematic the dominant discourses, values, and traditions of modernism, and by encouraging reflective analysis of, and moral deliberations over, the dilemmas of modernism as they relate to the profession of human movement, the authors of this book delineate ways and means for enhancing the probability of human agency and for facilitating social transformative actions that will create new egalitarian social structures, more humane conditions for everyone, and thus a better—a postmodern—world. The authors of the next chapters attempt to map out the role of human movement in this postmodern future.

CHAPTER 3

Gender Discrimination in the Norwegian Academia:
A Hidden Male Game or an Inspiration
for Postmodern Feminist Praxis?

Gerd von der Lippe

Introduction

My intentions for writing this chapter are to raise awareness
about masculine hegemonic orthodoxy and to denounce academic
practices that discriminate against women. At the outset, it is important to point out that the issues I will present here reflect my orientation as a woman, a feminist, a professor, and a political activist
(von der Lippe 1982, 1993). Although most of the issues I am about
to discuss will be framed within the context of Norwegian academia,
sympathetic readers may be able to extend them to other countries
and contexts.

Discourse, Difference, and Deconstruction

Among the postmodern terms feminists have appropriated in
their struggle for equity are "discourse," "difference," and "deconstruction." The term "discourse," as Foucault (1980a, 1980b) intended
it, refers to a specific structure of terms, categories, and beliefs—
based on historical, social, and institutional roots—that affects daily
life. In other words, a particular discourse gives a particular meaning to our lives. As such, meanings are contested within discursive
"fields of force." That is, the elaboration of meaning involves conflict
and power. Power to control a particular field resides in claims to
knowledge contained in social relations, language, and disciplinary

27

and professional organizations and institutions. Historically, these discursive fields have competed among themselves and those that have achieved more power have attempted to maintain their status by establishing their "truths" and "metanarratives" as legitimizing and self-evident. In turn, the rest of the discursive fields have become marginalized or silenced.

"Difference" is the notion that meaning is made through implicit or explicit binary oppositions, "that a positive definition rests on the negation or repression of something represented as antithetical to it" (Scott 1994, p. 285). For instance, in the modern era, difference in terms of gender has been tied to various types of cultural representations and relations in which men and women have been assigned different status and value not only as fixed oppositions, but also in ways that have accorded primacy and privilege to men, while viewing women as derivative and weaker.

"Deconstruction" is a process by which one analyzes the ways in which meanings are made to work (Culler 1982; Derrida 1976). Through deconstruction one can question the privileged status and legitimacy of established relations and forms of power and replace binary oppositions, dominant meanings, and unfounded functions. At the same time, the act of deconstruction enables one to bring to the foreground and assert marginalized meanings, histories, and texts.

Deconstructing Masculine Hegemonic Orthodoxy as a Discourse of Difference

In the modern era, masculine hegemonic orthodoxy has been one of the dominant discourses of difference. A *doxa* is a system of interdependent ideas used to explain and justify particular conditions, practices, and interests that seem to be "neutral" and self-evident. Over time, if a doxa is not submitted to debate and reflection, its meanings, practices, and interests are taken for granted and become uncritically accepted as part of the daily lifeworld. Put another way, the doxa becomes orthodoxy (Bourdieu 1991).

Oftentimes, orthodoxy serves the interests of the dominant groups of society and perpetuates and legitimates their ideas and practices projecting these into the mainstream of a culture. Once this occurs, it can be said that hegemony has been established (Gramsci 1971). Hegemony is "a social ascendancy achieved in a play of social forces that extends beyond contexts of brute power into the organizations of private life and cultural processes" (Cornell 1991, p. 184). In lay terms, hegemony means domination.

Regarding gender, masculine hegemonic orthodoxy is the result of certain dynamics by which, over time, masculine values, knowledges, and behaviors have gained power and privilege over those of women. Put simply, the notion that men are superior to women and therefore the former should have the "power over" and dominate the latter has become broadly accepted. Hegemonic orthodoxy (masculine or otherwise), however, is neither stable nor entirely successful. Indeed, it is a locus of struggle, and as such it should be subject to deconstruction.

In what follows, I will analyze and attempt to deconstruct masculine hegemonic orthodoxy. I will do so by analyzing (a) the barriers women professors encounter in the process of promotion and recognition in universities, and (b) the struggle to maintain a gender-equitable curriculum in physical education teacher education.

Setting the Stage

The competition for seemingly shrinking markets among countries of Western Europe, the U.S.A., Australia, and the prosperous parts of Asia has sharpened during the 1990s. Moreover, a fairly extreme market liberalism seems to dominate the philosophy behind projects such as GATT (General Agreement on Tariffs and Trade), the European Union, and NAFTA (North America Free Trade Association). In the midst of all this, it is not surprising that governments in Western Europe give priority to projects that can make their economic policies prevail and prosper. Lyotard (1984), in *The Postmodern Condition: A Report on Knowledge*, points out that "science seems more completely subordinated to the prevailing powers than ever before and, along with the new technologies, is in danger of becoming a major stake in their conflicts . . ." (p. 8). Hence, it is little wonder that knowledge production (scientific and nonscientific) is now perhaps more than ever geared toward the reinforcement of economic and political purposes, and thus takes priority over any other social issues. Likewise, because money for research at universities and other scientific institutions is still foremost granted within programs regulated by the State, "The question of the State becomes intimately entwined with that of scientific knowledge" (Lyotard 1984, p. 31).

The present economic pressures have had strong effects on the ways universities function and on the knowledges that are foregrounded or backgrounded within them. As governments struggle economically, their aid to universities has shrunk. In turn, in a period of shrinking budgets, universities have become more dependent on private donations, and these donations seldom come without strings attached. Private donations are supposed to be utilized to advance the donor's political and economic interests, and these interests are most frequently conservative. This makes it very unlikely for universities to argue for "particular" interests, and, consequently, feminists and other "radical" scholars (including their alternative values, knowledges, and agendas) are not only rendered less pertinent, but also perceived as intrusive and dangerous for the stability of the institution. Hence, "radical" and nontraditional scholars (e.g., women) are discriminated against and harassed (Clark et al, 1996) while their knowledge is marginalized and devalued.

Recently, several chilling reports have pointed to masculine hegemonic orthodoxy by denouncing the unsettling state of women within the professorial ranks in the U.S.A. and Australia. As a norm, women occupy the lower academic ranks, receive lower pay, have the heavier teaching loads, and hold most of the untenured, part-time, or adjunct positions. With regards to Australian higher education, Maslen (1995) asserts that "although the promotion of female faculty members has increased since 1988, the proportion of tenured professors who are women has fallen to 43 percent from 49 percent in that span. This is largely because most of the growth in faculty appointments has been in fixed-term or casual jobs, where women remain concentrated" (p. A43). On her part, Kolodny (1996b) has denounced the "exploitative" conditions of women's employment in universities: "[Women who are] part-timers and adjuncts are usually ineligible for health and retirement benefits; on some campuses they aren't even assigned an office. With large classes and mind-numbing teaching loads, fixed-term contact lecturers have little time to keep up with the latest developments in their field, and they are rarely reimbursed for the expenses of attending professional conferences. . . . These faculty members are hired and fired at will, whatever the quality of their job performance" (p. 24). Needless to say, under these circumstances, women have little power in the institution and, in order to keep their jobs, they must agree to teach whatever they are assigned and refrain from experimentation or innovation in their teaching practices, curricular content, and scholarship.

Masculine Hegemonic Orthodoxy in Academia:
Discrimination and Intellectual Harassment
of Women and Their Knowledge

Masculine hegemonic orthodoxy in academia reveals itself through the discrimination and intellectual harassment of groups of women and their knowledge. Kolodny (1996a, p. 9) has outlined the parameters of academic discrimination and antifeminist ideology and practices in what she calls "antifeminist intellectual harassment." As a serious threat to the rights of women and to academic freedom, *antifeminist intellectual harassment* occurs when any policy, action, statement, and/or behavior:

1. Has the effect of discouraging or preventing women's freedom of lawful action, freedom of thought, and freedom of expression;

2. Creates an environment in which the appropriate application of feminist theories or methodologies to research, scholarship, and teaching is devalued, discouraged, or altogether thwarted; or

3. Creates an environment in which research, scholarship, and teaching pertaining to women, gender, or gender inequities are devalued, discouraged, or altogether thwarted.

Antifeminist Intellectual Harassment in Tenure and
Promotion Practices within the Norwegian Academy

Within the European context, Norway is regarded as a country where the position of women is fairly good. "Only" about 4 percent of the female working force was unemployed in 1993 (Nei til EU 1994, p. 15), and the difference between men's and women's incomes is less than in most European countries. Further, we just had a woman Prime Minister, and, in three successive cabinets, females have held more than 40 percent of the seats (women were occupying 39 percent of the parliamentary seats in 1994). Moreover, the welfare state includes pensions for both the paid and unpaid work of women.

Paradoxically, these data do not hold true in academia, where the discourse is dominated mostly by men. In fact, not only the number of female professors across universities and regional tertiary colleges in Norway is quite low, but also women in academia are often discriminated against (Fürst 1988). In 1991, the number of positions of female academicians was rather low as compared to those of their male counterparts. Namely, 29 percent were female assistant professors, 16 percent were associate professors, and only 9 percent were full professors. Moreover, women in top administrative positions were only 2 percent.

In the past several years, Fürst (1988) and Holter (1991) have studied the discriminatory practices of male-dominated promotion committees at the University of Oslo regarding the promotion of women academicians. Through interviews with both women trying to achieve higher academic positions at this institution and with the members of several promotion committees, these investigators have gained some critical insights into the process of antifeminist intellectual harassment and discrimination.

In the promotion process, one of the more important aspects is the applicants' scientific and scholarly work. The criteria for judging this work, however, are formulated in such general terms that leaves the door open to many interpretations and manipulations. Here is one example of such criteria: "The emphasis [of the evaluation for promotion] is on the available scientific work, consideration given to experience. Leading of research projects and participation in these should also be included, provided that the work is adequately documented." As one can see, these are ambiguous criteria, with no clear definitions of scientific quality. This puts applicants who do not fit the patterns of traditional scientific work (i.e., feminist researchers and scholars) at a clear disadvantage and makes it very difficult for them to be promoted, especially if their research incudes questions about gender, a field about which most male professors seem to be ignorant and biased. The following quote from a male committee member with regards to his evaluation of a female applicant illustrates my point: "The applicant's great engagement in [gender] issues is now and then made very strongly. [However,] what this implies . . . [is] that she examines a problem which is important and interesting to her" (Fürst 1988, p. 98).

From this quote, one can see that the male evaluator did not deem gender issues as legitimate scientific discourse. If he had, it is reasonable to assume that, instead, he would have commented on the positive aspects of her research choice. This attitude seems to be a pattern within the Norwegian academic circles (i.e., the Parliamentary Commissioner for Equal Rights in Norway has received discrimination complaints in this regard from several women academicians across the country). Within these circles, feminist research seems to have a negative connotation and is often labeled as "a relative one-sidedness thematic" or "marginal" (Pedersen 1994, p. 13). Men's research, on the other hand, although usually more narrow in scope, is commonly described as "fundamental" and given positive evaluations. Not surprisingly, in the particular case just discussed, a male professor got the promotion. The rationale from the committee went as follows: "Although the work of [the female professor] has covered somewhat wider and more varied fields, [that of the male professor] is more concentrated in one field [and, therefore, more pertinent]."

Fürst (1988) argues that the elements of subjective evaluation by the male committee members make it easier for them to value more the qualifications of their male colleagues because evaluators are more familiar with these qualifications than with those of women. According to Haavind (1986), many male committee members do not seem to be aware of their incompetence regarding their lack of knowledge of feminist theories and women's work. In fact, many men who evaluate the scholarly activities of females do not read or understand women's research.

From this, one fundamental question emerges: Is feminist and women's research superfluous and self-righteous, or is the situation of women in academia worth reflecting about? I believe the latter. Moreover, another important question may be asked: What kind of discriminatory mechanisms are at work here? In order to answer this question, I must refer back to the concept of masculine hegemonic orthodoxy. Men are still the gatekeepers of important positions and create barriers to prevent women from reaching positions of power.

Implicitly, and to some extent also explicitly, the academic culture in Norway subscribes to the principle of the strong division between the person and her/his competence. According to Holter (1976), in their process of qualification, a lot of men seem to consciously or unconsciously use power techniques to devalue women. In this regard, Lorber (1993) found that one strategy male scholars use is to indirectly communicate negative messages or to make direct sexist remarks about female scholars and their work. Other strategies are silencing and rendering women "invisible." Moreover, women are forbidden from entering men's inner circles in the research world, where a lot of information is communicated in informal ways. If women were to enter men's circles and brought their own agendas and interpretations with them, quite a few men would feel pressured and uncomfortable—their power would no longer be exclusive.

Concerning Canada's academia, Dorothy Smith (1987) refers to these discriminatory practices as the establishment of a "male social universe." She claims that male scholars take the fundamental social and political structures for granted and utilize them as the point of departure for their power (Smith 1987). All this can be explained by the force of the masculine hegemonic orthodoxy. Under this discourse many men think of themselves as representatives of universal values while thinking of women as "bearers" of marginal ones. In this sense, women are portrayed as representatives of "not-as-relevant" *special* interests (Pedersen 1994), whereas men erect themselves as representatives of *general* interests (including those of women). Given the apparent "naturalness" of such a view, "Women are kept out of top positions [and assigned less important tasks] by sexism that is ingrained in men's attitudes and built into the structure of career mobility" (Lorber 1993, p. 64). Likewise, Harding (1987) points out that, in order to complete their scholarly projects, many male professors have female assistants and wives do the daily practical work.

Antifeminist Intellectual Harassment in Physical Education Teacher Education Curriculum

This section also deals with both orthodox masculine hegemony and antifeminist intellectual harassment in academia. In this case, I will share my lived experiences regarding gender issues in physical education teacher education curriculum in Norway. Here, again, some informational background is called for. In Norway, there is a National Committee of Physical Education and Sport in which fourteen different institutions of higher education are represented. Four of these institutions offer graduate studies in physical education (including ours at Telemark), one offers doctoral studies (i.e., Oslo), and the rest offer one- to three-year degrees in this field.

Despite the presence of all these institutions in this committee, not all of them have the right to vote. Hence, from the start, this committee is not fully democratic. Moreover, women's voices are also severely curtailed because all but one of the members of this committee are males (this constituency will remain so until 1996).[1] And this is not all. The only female member in this committee is a student, and, obviously, her status grants her very little power.

One of the functions of this committee is to develop new curricula for the studies in physical education and sport. Since 1992, the committee has been discussing whether the initial studies in different institutions should have similar curricula so that the students may easily switch from one institution to another without jeopardizing their chances of being admitted into graduate programs later on. In particular, one of the issues debated more strongly is whether students should have a similar curriculum in psychology in their first year of study. The point of reference in the committee's discussion is the content in psychology at the Norges Idrettshøgskole (NIH) of Oslo. Traditionally, a lot of students from other districts have continued their education at this central institute, and, therefore, facilitating their transfer there has become an important matter for the committee. Although this is a legitimate concern, what poses a problem is that the discussions about this common curriculum have focused on pragmatic factors only. Other factors such as the educational costs of this curricular change for students have not been seriously contemplated.

At our institution in Telemark, one of the primary goals is to educate students critically, both in theory and in practice. With this goal in mind, we teach our curricular subjects in an integrated way, combining several areas (e.g., history, psychology, sociology, and pedagogy). We center lessons around holistic themes such as "Freud and gender perspectives," "hegemonic attitudes affecting the teacher's role in elementary physical education," and "sport and society." This curricular orientation has a hermeneutic (not a positivistic) orientation and includes traditional as well as feminist and critical perspectives. If the proposed unified curriculum were to be implemented, we, at Telemark, would have to dramatically change our pedagogical philosophy. Also, and perhaps more important, we believe that this change would deny a more sound and equitable preparation to the majority of our students at Telemark College. In fact, only about a third of our first-year students (we presently have seventy-six) normally transfer to the NIH in Oslo.

Notwithstanding our arguments, the leader of the National Committee has pointed out that psychological theory is a subject which has nothing to do with feminist and critical perspectives. According to him, psychology is supposed to be gender-neutral and objective. His opinion is that psychological theory must be taught in a vacuum, disregarding the complex social and political aspects implicit in this subject. Unfortunately for us, the majority of the members of this committee agreed with the chairman and voted to implement the new curriculum. This is another example of the ignorance on the part of male professors about gender and other related issues.

As of this writing, the situation is as follows: With the intention to raise public and academic awareness about this curricular matter, and with the hope to force a reversal of the committee's decision, I wrote and published a short article explaining the situation and presenting our arguments. Furthermore, our institution followed legal procedures. In Norway, clause No. 21 of the Law on Equal Rights deals with gender quotas in official groups. This Law establishes 40 percent as a minimum quota in terms of gender membership for committees such as this one. This law helped us question on legal grounds the unequal composition of the committee. Moreover, we also networked with, and secured the support of, several Norwegian physical education centers of higher education. Our actions turned to be fruitful, and action on the committee's decision about the core curriculum in psychology was postponed. The struggle, however, continues.

Closing Remarks

Some argue that postmodernists have focused primarily on philosophical perspectives (Fraser and Nicholson 1990). Barrett (1992) puts it this way: "In terms of aesthetic strategies and cultural forms, postmodernism refers to an interest in surface rather than depth . . ." (p. 206). I do not agree with this interpretation because it refers to theory only, leaving practice and reflection aside. I, like Benhabib (1994), believe that feminist emancipatory theories have little sense unless feminists engage in praxis. Praxis—the cyclical process of reflection-action in order to change the world—is what matters here (Freire 1970), because with praxis come the intentionality, the responsibility, and the power to struggle for what is right and just. That is why the relationship between theory and practice is important to feminists—it is through praxis that we gain the strength and the insight to argue and struggle for issues such as the fair treatment of female professors and for gender equitable curricula. Academia in the twenty-first century may benefit from the lessons and praxis of the 1970s, when cooperation among feminists widened the scope of knowledge and created more proportional opportunities for women and men alike.

I believe in the capacity of human beings to dialogue, reflect, and act in order to change our world and bring about more justice. We must do so in order to promote acceptance of alternative ideas and the sense of a human community (Bourdieu 1991). We can change things and we must act on that premise. If not, life would be reduced to the past, and its meaning will be determined by the already-gone-by, unchangeable history, not by the possibility of a better future.

I, for one, am committed to advocating women's' rights. However, I consider it impossible to struggle for all women's rights all the time. A strategy that works for me is to identify some key questions and issues with which I can deal and about which I can make a difference. Then, through alliances with other groups of women and men, and by making compromises to include as many people as possible, I attempt to bring about change in the traditional values and practices of masculine hegemonic orthodoxy. It is in this way that the postmodern discourse continues for me.

CHAPTER 4

Schooling Bodies in New Times:
The Reform of School Physical Education
in High Modernity

David Kirk

Introduction

The central purposes of this chapter are to illustrate some of the ways in which school physical education and sport have been both participants in and outcomes of the processes of constructing and constituting the body in modernity, and then to examine the shifts in physical education's treatment of the body during this century as indicative of shifts in aspects of popular physical culture more broadly.

Physical education in Australian schools has been scrutinised recently by delegates attending a national workshop of physical educators ("Australian Physical Education in Crisis," Deakin University, October 1991), by a prominent television current affairs team (*Four Corners*, October 1992), and by a senate inquiry into physical and sport education which reported in December 1992. The Senate Inquiry Report reflected the critical tone of these reviews of physical education in its summary statement, which identified widespread support for physical education and sport but a dramatic reduction in the quality and quantity of physical education provision in schools. The report suggested that school physical education and sport are in serious decline in the early nineties, with schools struggling to offer programs in the face of shrinking resources, high teacher attrition, and growing disinterest on the part of students (PCA 1992).

Revealing (somewhat inadvertently) the symbolic power of the performing, sporting body, the sport journalist Jeff Wells (1990) has seen in this apparent decline a metaphor for the moral bankruptcy of Australian society writ large, suggesting that it is nothing less than criminal to permit the neglect of children's bodies since such a neglect undermines a way of life which is sacrosanct, for "Aussies" in all parts of the world are doing a vigorous job at something. Wells's linking of physical education and sport to matters of economic performance and national decline is not new (Hargreaves 1986), though his lack of originality makes this articulation no less significant.

Physical education and sport in schools take as their task the shaping of children's bodies, both biologically and socially. Given this task, it is in some respects surprising that we seem to be witnessing a decline in the fortunes of school physical education and sport when the body and bodily practices such as sport, exercising, and dieting are taking on an increasingly significant role in high modernity.[1] In this chapter, I want to argue that the socially constructed body and physical culture[2] in toto are of central importance to understanding the nature of new times, since the body in culture is both a surface reflecting and refracting cultural change, and an active agent in producing change. In particular, I am interested in exploring the shifting position of the body in modernity through the refracted lense of school physical activity programs and assessing in light of these historical lessons possible ways forward for physical education in new times.

The growing cultural importance of the body has not gone unnoticed by sociologists in contemporary Western societies (e.g., Turner 1984; Turner 1992; Franks 1990), though few have sought explicitly to pursue the links between bodily practices in schools and those carried out in other sites. Shilling (1991) is one of the few sociologists who has begun to explore these links, and he does so by drawing substantially on the notion of *habitus* (Bourdieu and Passeron 1977), which Shilling claims is located within the body, every gesture and physical act betraying an individual's class-dependent social location and orientation to the world. Shilling has shown how the body and the physical capital invested in it play key roles in the production of social inequalities, and how school physical education and sport contribute significantly to this process. Pointing out the class and gender dimensions of these inequalities, Shilling suggests that physical capital can be turned to economic and social gain, but social class location delimits the range of opportunities available for such conversion (Sage, chapter 2). Clearly thinking of activities such as rugby union, polo, and lacrosse, Shilling claims that "children from (the dominant) classes tend to engage in socially elite sporting activities which stress manners and deportment and hence facilitate the future acquisition of social and cultural capital" (Shilling 1991, p. 656).

In his recent comprehensive overview of a range of sociological writing dealing with the body, Shilling (1993a) argues that the social construction of bodies needs to be located within particular social locations, the *habitus*, and through the development of taste. He argues that, for growing numbers of people from a range of social class, age, and ethnic groups, the body has become an individual project in the 1990s, an issue which is discussed in relation to exercise by Tinning (chapter 7), Fahlberg and Fahlberg (chapter 5), and Ingham (chapter 10) in this volume. Shilling's argument is underpinned by two propositions. The first is the widespread and commonplace acceptance of the idea that the body is malleable, a view supported by growing knowledge and technical expertise of means of intervening in and substantially altering the shape and look of bodies. Since the 1970s, as Tinning demonstrates (chapter 7), there has been fast emerging from Australian tertiary institutions a cadre of sport and exercise science professionals who claim to be "body maintenance experts" (McKay 1991) able to apply this new knowledge. The second is a growing awareness of the body as an unfinished project which can be pursued to some kind of resolution according to the life-style choices people make. Drawing selectively on the work of P. Bourdieu, A. Giddens, B. S. Turner, and others, Shilling remarks that contemporary preoccupation with the body-as-a-project is set apart from various premodern treatments of the body by the conflation of physical appearance and self-identity, wherein the body is the flesh-and-blood manifestation of social and, more recently, individual identity.

Foucault (1980a) captures something of the nature of these shifts suggesting that social regulation of the body in the mid to late twentieth century is accomplished through distinctively different processes to those of the preceding two centuries. He argues that

> from the eighteenth century to the early twentieth century I think it was believed that the investment of the body by power had to be heavy, ponderous, meticulous and constant. . . . And then, starting from the 1960s, it began to be realised that such a cumbersome form of power was no longer as indispensable as had been thought and that industrial societies could content themselves with a looser form of power over the body. (p. 58)

Foucault's characterisation of this shift in the manner of normalisation and regulation of the body offers a useful focus for the argument to be pursued in this chapter, since it is suggestive of a body in high modernity, which is regulated less coercively and externally, by others, and more often internally by the self. The shift to a looser form of corporeal power suggests an associated shift in the locus of control, from external and mass practices to internal and individual practices, and offers some support for Shilling's (1993a) suggestion that the body has become an unfinished, individualised project in the 1980s and 1990s.

The apparent decline of school physical education and sport programs, in the face of the growing prominence of concerns for body maintenance (Featherstone 1982), provides an opportunity to examine some dimensions of the diffusion, individualisation, and internalisation of corporeal power in new times.

Schooling Docile Bodies: Heavy, Ponderous, Meticulous, and Constant Power

The emergence of various systems of rational gymnastics towards the end of the eighteenth century (Munrow 1955), and their eventual widespread adoption by a number of institutions such as schools and the military by the end of the nineteenth century, was a constituent part of the development of a range of regulative and normative practices aimed at schooling a docile body. As Foucault (1977) has observed, docility did not imply subjugation of the body, since economic productivity was partnered with the effective use of the labour of compliant and healthy citizens. On the contrary, in Foucault's terms, the "little practices" of schooling the body were meant to achieve "docility-utility," without the need for routine violent punishment (Kirk and Spiller 1994). Even in institutions like the British navy, where flogging continued to be a regular means of punishing sailors into the mid-1800s, a system of discipline emerged towards the end of the eighteenth century which attempted to secure the willing compliance of seamen and make more efficient use of their labour (Dening 1993).

Through the ritualisation of practices such as flogging and the precise codification of discipline, officers and crew were part of a web of authoritative power relations in which the system, rather than any individual, disciplined an offender. Significantly, the British navy was just one institution which further developed this approach to discipline through the use of the Swedish system of rational gymnastics from the mid-1800s (McIntosh 1968), in which heavy, ponderous, and meticulous practices of schooling a docile body were applied. Australian government schools, as we shall see, were to adopt the same (Swedish) system of drilling and exercising early in the 1900s.

From their first appearance in the mid-1800s, physical activities in Australian schools can be viewed as practices of corporeal regulation and normalisation which were integral to the opening and operation of at least two institutions of modernity, surveillance—or "the control of information and social supervision"—and capitalism—or "capital accumulation in the context of competitive labour and product markets"—(Giddens 1990, p. 59). As Turner (1984, p. 161) has argued, following Foucault, from the early 1800s "capital could profit from the accumulation of men and the enlargement of markets only when the health and docility of the population had been made possible by a network of regulations and controls."

This entanglement of corporeal regulation and normalisation with surveillance and capitalism was part of a process of reifying and constructing the modern body which had been in train since the Renaissance, through which the body was increasingly identified with "personhood" (Broekhoff 1972). Physical appearance was conflated with self worth or value (Finkelstein 1991), initially defined as a classed, raced, and gendered (or socially positioned) self, and later (and additionally) as a self-as-individual. S. J. Gould (1981) has shown how craniometry, criminal anthropology, and mental testing in the 1800s and the early 1900s each attempted to legitimate "scientifically" this conflation of the social worth of individuals and their physical appearance, particularly in relation to "proving" the intellectual and moral inferiority of Blacks and women to white males. By the end of the nineteenth century, we can see these notions being worked through in a range of mass, corporeal, regulative, and normative practices described by Foucault (1977), constituting prisons, schools, factories, and barracks. It is here, within this nexus of practices, that early forms of bodily practices in Australian schools can be located, particularly drilling and exercising (Kirk and Twigg 1993), school medical inspection (Kirk and Twigg 1994), and competitive team games (Kirk and Twigg 1995). Each was a site for the surveillance of bodies as they were shaped to meet particular social and economic ends.

The emergence of drilling and exercising, school medical inspection, and competitive teams games in Australian schools exemplifies in relatively highly codified forms the notion that the body is not a purely "natural" phenomenon, despite the hegemony of medical and biological science, but is also a cultural aspect and can be normalised and regulated according to particular social, economic, and cultural purposes. By the end of the nineteenth century in Australia and elsewhere, the use of forms of physical activity for the purposes of shaping a docile body within the context of a particular *habitus* was explicitly prescribed in the official discourse of educational policymakers, manual writers, and head teachers of elite schools. There followed a period of consolidation of these school practices during the first three decades of this century, through the institutionalisation of a drilling and exercising form of physical training in governmental schools, of school medical inspection, and of games-playing in government and non-governmental schools.

The processes of body shaping had two key features during this period. The first of these was that children were usually treated in the mass rather than as individuals. The second was that these practices of corporeal regulation and normalisation relied in the main on securing children's compliant participation through the enforcement by teachers and other adults of precise and meticulous prescriptions and measurements detailed in texts and manuals, in the cases of drilling and exercising and medical inspection, or in the case of games-playing, through the strict application of an unwritten but all pervasive code of gentlemanly or lady-like conduct.

Drilling and Exercising As Codifications of Corporeal Power

An example of the codification of ponderous and meticulous corporeal power aimed at schooling docile bodies is the system of physical training in operation in Australian schools between 1911 and 1931, conducted under the auspices of the Commonwealth Department of Defence in association with the Junior Cadet Training Scheme (Kirk and Twigg 1993). While only twelve to fourteen-year-old-male pupils were eligible to become junior cadets, Defence Department instructors ran courses for teachers based on the *Junior Cadet Training Manual*, published in 1916 by the Australian Military Forces, and so had a more or less direct influence on the physical training of girls and younger boys. Lessons in the *Junior Cadet Training Manual* drew on the Swedish system of gymnastics and were scripted in the form of tables which dealt, in systematic fashion, with the major joints and muscle groups of the body. The tables themselves were organised in an immutable sequence according to the age and experience of the pupils. This amounted to an entire physical training program, with progressions, sequencing, and age standards mapped out in detail (table 4.1).

Table 4.1
Sequence of Exercise Tables According to
Age and Experience of Pupils

Series	Approximate Age of Pupils	Years of Work in Physical Exercises	Terms: First	Second	Third
A	7–8	First	1–4	5–8	9–12
	8–9	Second	13–16	17–20	21–24
B	9–10	Third	25–28	29–32	33–36
	10–11	Fourth	37–40	41–44	45–48
C	11–12	Fifth	49–52	53–56	57–60
	12–14	Sixth	61–64	65–68	69–72

Source: Australian Military Forces (1916). Junior Cadet Training Manual. Melbourne: Government Printer.

The lessons themselves were set out in a format which also had to be adhered to strictly. There were eight categories of activities, and each lesson was constructed on the basis of a selection of exercises from each category. The categories were, in order of appearance in the lesson: (a) introduction and breathing exercises, (b) trunk bending back and forward, (c) arm bending and stretching, (d) balance exercises, (e) shoulder-blade exercises (abdominal exercises), (f) trunk turning and bending sideways, (g) marching, running, jumping, games, and so forth, and (h) breathing exercises. Teachers were required to memorise the precise series of exercises for each lesson, and to deliver their instructions using such commands as "head backwards . . . bend!" "left foot sideways . . . place!" "trunk forward and downward . . . stretch!" and "knees . . . bend!" These commands were amended to "suit Australian conditions" in 1922 and again in 1926, but the militaristic flavour persisted. Lesson 2 in the sequence outlined in the 1922 edition of the *Junior Cadet Training Textbook* (Department of Defence 1922) begins in the following way:

1. Exercise description: Free running in large circle, instant halt on signal.
   Command: "Double . . . march! Class . . . halt!"

Exercise description: Run to form one rank at wall, place leaders on marks, run to open ranks.

Command: "Back to the wall . . . move! Leaders on markers . . . move! To your places . . . move!"

2. Exercise description: Astride, trunk bending downward to grasp ankles.

Command: "With a jump, feet astride . . . place! Grasping both ankles . . . down! Class . . . up! With a jump, feet together . . . place!"

This same formula was still in use in the 1930s, even after the Defence Department had retired from the scene, and on into the 1940s. Correctness of performance was the overriding principle in this approach. Teachers were advised by Rosalie Virtue, the Victorian Education Department's Organiser of Physical Training between 1915–1938, that "correct starting positions and the correct performance of exercises are essential" and that "quickeners" should be put in to keep the children mentally alert. "Teach only one new exercise each day," she stressed, because "this allows repetition to obtain correctness of performance, a gradual change of table, and more work done by children" (*Education Gazette*, Victoria 1933).

The scheme of physical training embodied in the various physical training manuals was, effectively, the first and last national program of physical activity in Australian schools, though its implementation was dogged by problems from the start. World War I caused massive disruptions due to the posting of servicemen overseas. These disruptions encouraged criticism from commentators who, as early as 1917, were referring to the "tedious monotony of elementary drill," suggesting that this formula for teaching physical training lessons was more likely to cause resentment rather than foster willing compliance, since it relied entirely on the teachers' strict adherence to the prescriptions of the manual and left no room for initiative or expression on the part of either pupils or teachers (*Argus*, April 1917). For two years in the early 1920s, the scheme was temporarily abandoned, only to be resurrected on a reduced scale following vehement protests from the state governments.

While the State Education Departments resented the involvement of the military in their schools, the attraction of a scheme of physical training funded by the Commonwealth which, ostensibly, benefited not only junior cadets but all children, outweighed their concerns. The Department of Defence was finally forced to withdraw its services in 1931, again under protest from the states, when the scheme was suspended in 1929 by the Scullin Labor Government in a climate of deepening economic and social crises (Kirk and Twigg 1993).

This rather chequered career between 1911 and 1931 meant that the implementation of this drilling and exercising form of physical training was uneven across the country and could not have been wholly effective. Nevertheless, these pedagogical practices in schools, appearing as they did at a time when questions of national and racial identity had a potent influence on public and professional discourses, reveal highly codified and institutionalised attempts to normalise and regulate children's bodies, docile bodies which were both compliant and productive. Regardless of the actual effectiveness of drilling and exercising, their sociological significance lies in their use as a strategy of corporeal power, focusing in this case specifically on the construction of acquiescent and productive working-class bodies.

Linked to school medical inspection, which was inaugurated between 1900 and 1910 in most states and which sought to map and measure an entire range of physical "defects" among students (Kirk and Twigg 1994), drilling and exercising focused, in a ponderous and meticulous fashion, on the movements of children's bodies in space and time. This form of physical training was intended to codify corporeal power through its precise definitions of physically appropriate and inappropriate activity, its shaping how and where the body might move, and the range of outcomes physical activities might produce.

The Invention of "Physical Education" and the Emergence of a Looser Form of Power

The 1940s marked a sea-change in this process of corporeal regulation and normalisation within schools, with a gradual shift from treating children's bodies in the mass to a greater concern for individual bodies and for less regimented and strictly prescribed forms of movement. Hence, the reformist agenda of inspection and intervention proposed by school medical officers in the pre-first world war period, informed largely by a philosophy of positive eugenics, was watered down somewhat during the 1920s in the aftermath of the war as radical eugenicists advocated drastic solutions to the problems of feeblemindedness and other forms of "abnormality," though the same strategies of medical inspection and the identification of "defectives" continued to be employed well into the late 1930s. After this point, in the early 1940s, medical inspection's link to physical activity programs in schools was loosened by the relocation of the School Medical Services from state education to health departments (Kirk and Twigg 1994). This development reflected broader changes in the ways in which bodies were treated within government schools, with the emergence in the late 1930s of a new form of physical education as a comprehensive program of physical activities that had competitive team games at its core.

Following its successful passage through Parliament in 1941, the National Fitness Act was instrumental in establishing three-year-diploma courses for specialist physical education teachers in most of Australia's universities, a development which heralded the emergence of a homegrown physical education profession. It was also during this period that a definition of "physical education" began to be constructed out of a cluster of disparate physical activities. By 1929 in Victoria, the Minister of Education in his annual report suggested that physical education "includes not merely formal physical exercises, but swimming, organised games, rhythmic exercises, folk dancing, practical hygiene, and remedial exercises based on the medical assessment of the needs of each child" (*Vic. MPI Rpts.*, 1928–1929, p. 8). This notion, that the term "physical education" might embrace this comprehensive range of practices, was new at this time and not widely accepted nor understood in the education community. There was, for instance, no subject called physical education in schools beyond drilling and exercising lessons. Competitive team games were to exert the greatest influence over the new notion of physical education as a comprehensive program of physical activities for schools.

Games were already well established in government schools by the late 1920s, with state schools' amateur athletic associations being formed around 1906 in Victoria and New South Wales to organise interschool and interstate matches. Only the major team games were played, plus athletics and competitions in swimming, and these activities were extracurricular. There was no instruction in games-playing within curriculum time, though there was certainly some coaching by teachers of teams which competed in interschool matches (Kirk and Twigg 1995). By all accounts, competition was taken very seriously and, significantly, was continually justified by appeals to what Mangan (1986) has described as the "games ethic."

Children in the schools serving Australia's social elites had been participating in organised sport competitions since at least the 1880s, and this practice was embedded in versions of the English Public Schools games ethic. In the context of these elite English schools, games-playing was integral to constructing Christian manliness and fostering leadership, and was in itself a form of corporeal normalisation and regulation suited to the particular *habitus* of the middle and ruling classes (Sherington 1983). This ethic was adopted with only minor modifications in elite schools for girls (Crawford 1984).

Crawford (1981) has argued that, over time, the bourgeois English version of the games ethic was reconstructed in the antipodes to provide a distinctively Australian flavour. This reconstruction of the games ethic in Australia's elite schools may have reflected, as Crawford has claimed, some indigenous qualities of Australian middle class manhood. At the same time, games-playing as a form of leadership education was not a key concern in the education of the working classes, though it may have been viewed as a means of moral improvement and of civilising the working-class body. Notions of manliness were prominent in early justifications of games for boys in government schools.

According to Crawford, following the carnage of World War I, there was a pronounced emphasis on physical development through games and other physical activities in an attempt to repair "the national physique." Later, through the 1920s, we find greater attention being paid to values such as cooperation, courage, and playing for the sake of the team, as much as means of counteracting undesirable behaviour, like cheating, than as positive virtues in themselves. Later still, in the 1930s and 1940s, under the influence of the progressive movement in primary school education and perhaps in more conscious consideration of girls in addition to boys, concepts such as self-confidence, enjoyment, and play began to be added to the list of positive qualities games were claimed to foster.

This reconstruction of the games ethic for use in government schools can be seen in an important series of articles which appeared in a 1941 issue of the Victorian *Education Gazatte* (Hamilton 1941). The articles addressed the problem of the "cult of athleticism" which had plagued sports contests among the elite schools, involving boisterous and sometimes violent behaviour by players and spectators alike. In similar fashion, excessive zeal and ferocious competition were not uncommon in sports contests among government schools. It was advocated, in contrast, that games are the means by which every child can be given an interest in physical activity. Children should be taught to gain satisfaction from seeing their own improvement in performance, and not necessarily from competing.

Appeals of this sort, that higher qualities be developed through games playing, were commonplace in official government policy statements throughout the 1920s and 1930s. However, the notion that games could benefit the majority of children, not merely the socially privileged or physically talented, added a new twist which ran counter to some of the earlier elitist connotations of the games ethic. By the end of World War II, this view that games-playing was the generative force behind a definition of "physical education" as a comprehensive program of educational physical activities, and sport, in the form of major team games, was firmly lodged at the heart of this notion of physical education.

In 1946, a textbook for physical education prepared for use in Victorian schools, known as *The Grey Book*, crystalised this new concept of physical education (Education Department of Victoria 1946). In the foreword to this new Victorian textbook, comprehensive physical education was contrasted with the drilling and exercising form of physical training *The Grey Book* sought to displace, arguing that "formal exercises are artificial, unrelated to life situations, and generally lacking in interest; they also completely ignore the strong influence emotions exert on the physical well-being of the individual" (p. xii). The writer went on to map out the key dimensions of this definition of physical education which are still current today. "Enjoyment and enthusiasm" were recognised as of central importance to the beneficial effects of participation in physical activity, in contrast to the formality of the former regime of drilling and exercising.

The influence of the progressive, child-centred movement in primary education is also strongly in evidence. Adopting the notion that play is a natural activity for children, the writers of *The Grey Book* commented that "every child has the right to play, and this right must be restored to all children who have lost it" (p. vi). In a significant conceptual leap, the writers went on to equate "play," within this new notion of physical education, with playing competitive team games, and in one stroke conflated the key elements of the middle class games ethic with the progressivist's notion of play, through the idea that all children in government schools had the right to participate in competitive team games. This probably continues to be the single most significant conjoining of concepts underpinning contemporary physical education programs in Australia, since it positioned sport as pivotal to the educational legitimation of physical education. Few other conceptualisations have the symbolic power of this particular discursive configuration and, indeed, it is precisely by virtue of the invention of this notion of physical education with competitive team games at its centre that Jeff Wells has been able to make his equation of the decline of physical education in Australia.

One other relationship remained to be constructed at this time, which was the nexus between the specialist physical education teacher's role in time-tabled lessons and competitive sport in schools. With growing numbers of specialist advisers in the primary school sector and specialist physical education teachers in the secondary schools during the 1940s, 1950s, and 1960s, it became possible to provide instruction for all children in the skills considered to be prerequisite to games playing. The specialists took on this role with considerable enthusiasm. In so doing, physical education had to be conceptualised as the base of a pyramidal structure which had elite sport competition at the top. The majority of children participated in school physical education, while only a few talented individuals survived to reach the pinnacle of the pyramid (Evans 1990). Physical education lessons provided the "fundamental motor skills" of running, throwing, jumping, kicking, and so on, and these were then applied within an ascending scale of competitive contexts in interschool, interdistrict, interstate and international sport.

In a June 1993 edition of an *Australian Council for Health, Physical Education and Recreation Newsletter*, the Director of the Victorian Institute of Sport, Frank Pyke, articulated precisely this relationship between school physical education and competitive sport in an article titled "No Base, No Pinnacle." Responding to the current decline of physical education, he argued that "Australia needs tennis players who can not only hit the good shots, but can run down the returns with speed, agility and fitness. We need cricketers who can make the extra run when batting and prevent the extra one when fielding because they can sprint, dive, tumble, and throw well . . . the foundation for all of these skills is provided in the primary schools where young minds and bodies are ideally suited to rapid and efficient learning" (Pyke 1993, p. 8). It was during the watershed period of the 1940s that this particular contemporary notion of "physical education" was invented and began to be positioned as the "foundation stone" for children's participation in sport, as the site in which the skills required for sports participation should be developed.

Comprehensive Physical Education:
A Looser Form of Power?

The invention of physical education as a comprehensive program of physical activities, which foregrounded sport-related skills and formed the foundation for competitive sport, and its implementation in schools after World War II, was part of a liberalising movement in primary school education. From the earliest forms of physical activities in schools in the late 1800s to the arrival of comprehensive physical education in the 1940s, we can see marked contrasts in the ways in which anticipated outcomes were expressed. For instance, the Victorian Inspectors of Drill argued in 1889 that "with a compulsory system of drill, incipient larrikinism would receive a severe check, and the military spirit of the colony would be greatly fostered" (*Vic. MPI Rpts.*, 1889–1890, p. 264), while some sixty years later the writer of the foreword to *The Grey Book* stressed the "right of all children to play" alongside with "enjoyment" and "well-being." The shift in corporeal power signaled by these changes was further elaborated in school programs over a forty-year period between the 1940s and the 1970s. The emphasis changed from treating the mass of bodies to the individual body (evidenced in new teaching methods of individualised skill and fitness development) and from external prescription and enforcement to internal motivation to participate (evidenced in the concern for children's enjoyment of physical activities and the development of positive attitudes and lifelong participation).

Notwithstanding this shift and these new foregroundings, the institutional imperatives of schooling continued to demand that children be taught the same material in age-graded classes, delimiting the implementation of individualised approaches to students' learning which the liberation from drilling and exercising seemed to herald. While there may have developed a looser form of power over bodies in physical education and sport lessons in the post-World War II period, in many secondary schools, physical educators continue to be perceived within the wider school community as enforcers of discipline through semiregimented, command style teaching methods (Woods 1979). Moreover, as team games increasingly have become the dominating feature of school physical education programs, the individualism of liberal progressivism coexists in uneasy tension with the need for sublimating individual needs to those of the group or, in this case, the team.

Contemporary physical education and sport in schools has only recently been forced to come to terms with this legacy of corporeal normalisation and regulation, brought about mainly by an acceleration of interest during the 1970s and 1980s in body management within the population at large. Current treatment of the body in schools continues to frame the mass practices of developing sport skills and physical fitness within a liberal humanist philosophy of enjoyment, choice, and lifelong participation. In the last decade, we have witnessed the widespread adoption by physical educators of notions such as "active lifestyle" as key goals of their programs, mirroring developments in popular physical culture more broadly. While the pedagogy and organisation of school programs differ little from the practices of the immediate postwar period in terms of treating children's bodies in uniform ways, the range of activities now available in some schools has broadened considerably, and pupils have in many secondary schools greater choice of activity than ever before. These developments, as logical elaborations of the discourse of comprehensive physical education, are clearly less prescriptive and meticulous in their treatment of children's bodies in time and space as their forerunners, drilling and exercising: they seem to represent a looser form of power over the body, even though their operation is circumscribed by, and significantly contributes to, the institutional priorities of schooling.

New Bodies for New Times

However, new regulative and normative practices associated with the representation and management of the body, which take the malleability of the body and the conflation of the body with self-identity well beyond the means available to physical education and sport, have brought into question the continuing cultural relevance of these school practices. The acceleration of interest in body management as an unfinished project predicated on life-style choices has diffused, internalised, and individualised corporeal power much more fully and profoundly than physical education in schools. Since the late 1970s, there has been an increasing visibility of the body in popular cultural forms such as the television advertisement, revealing the extent to which the body itself has become a commodity (Fitzclarence 1990). It is now widely acknowledged that in most post-industrial societies young people's exposure to visual media has increased substantially in the past two decades (Fiske 1987), a development which has stimulated considerable debate over the effects of all forms of visual media on a range of social behaviours as well as on cognitive and affective development (Aronowitz and Giroux 1985; Postman 1985; Winn 1977), while feminist scholars have been critical of the perceived links between sport-based physical education and male violence (e.g., Waite 1985).

    In train with these developments, the relationship between sport and visual media has changed. The ongoing commercialisation of

sport and technological developments such as satellite and cable television have begun to alter the form and content of sports contests and
the experience of the spectator viewing these contests (Gruneau
1988; McKay and Rowe 1987). Media sport, that recent hybrid of live
sports contests and television, has greatly added to the length of time
bodies are exposed to the viewer and voyeur. A related development
has been the increasing prominence of sporting and exercising bodies
appearing in the form of advertising and other commercial and pro-
motional activities. It is significant that body shape is often the pivotal point of focus in contemporary corporeal discourse since it is
visual media, in both its televisual and popular magazine forms,
which have been the main conduits for transmitting images and representations of the body. The "normal" body for many women in Australia and elsewhere in the West is now much more slender than it
was in the 1950s, and the conjunction of representations of bodies
with the consumption of products (through advertising) has created
desire for corporeal normality and a consequent willingness to submit
to self-imposed regulatory regimes such as dieting and exercising.

Shilling (1993a) has suggested that the prominence of bodies in
all forms of visual media is predicated on a shift in understanding
that bodies are generators as well as receptors of social meanings and
relationships. Rothfield (1986) points out that the renaissance view of
bodies as natural objects has tended to make it difficult to think
about bodies in other ways, especially in terms of the relationships of
the body to social and cultural processes. This relatively recent shift
in understanding has resulted in a view of bodies as socially constructed, as existing in culture as well as in nature (Kirk 1993).

All bodies signify in all aspects of everyday life (Ingham, chapter 10). A good example of this process of signification is the values
which are associated with various body shapes and sizes. Bodies are
socially constructed through processes of learning and interaction,
so that acceptable facial and other physical gestures, comportment,
body adornment and decoration, and a host of other physical acts, as
these are defined by particular groups and communities, are appropriated for use in everyday interaction (Mauss 1973). Children learn
about their bodies and their capabilities in both physical and social
environments, in relation to objects and other people. Bodies are the
practical mode of engagement with a range of external events and
situations, and it is through relentless and continuous monitoring of
bodies and their expressive capabilities that individuals successfully
engage in social activity (Giddens 1991). Researchers interested in
body-size dissatisfaction have shown that the social consequences of

perceiving oneself to be overweight impacts negatively on feelings of self-esteem and can contribute to eating disorders, particularly among women (Tiggeman and Pennington 1990).

In consumption-driven societies, the everyday signifying processes in which bodies play a central part take on new associations through the commodification of bodies (Fitzclarence 1990). The social consequences of body commodification take at least two forms, both of which objectify bodies. The first form incorporates bodies within the cycle of consumption, involving an increased concern for what Featherstone (1982) calls body maintenance, where bodies "require servicing, regular care and attention to preserve maximum efficiency" (p. 24). In this respect, bodies become a site of consumption in themselves, particularly of the services of the beauty, cosmetics, clothing, exercise, and leisure industries. In the second form, bodies are a focal point of the commercial process, and the everyday signifying properties of bodies are rearticulated to make new associations with particular commercial products. Bodies are used to sell products by linking these products to particular significations such as mesomorphic slenderness as a symbol of fitness, health, and social success.

In this regard, Bordo (1990) has argued that slenderness is the dominant desired body shape in contemporary Western cultures, particularly among women. Signaling the gradual internalisation of surveillance, she suggests that the slender body is a manifestation of normalising strategies which ensure "the production of self-monitoring and self-disciplining docile bodies, sensitive to any departure from social norms, and habituated to self-improvement and transformation in the service of these norms" (p. 85). However, also according to Bordo, the looser form of power mentioned by Foucault does not necessarily mean greater freedom for individuals. Since the quest for slenderness marks an internalisation of the process of social regulation, Bordo is interested in representations of the slender body as analogous to the social body and as a symbol of power and power relations, especially among and between men and women. She suggests:

> The "correct" management of desire in (consumer) culture, requiring as it does a contradictory double-bind construction of personality, inevitably produces an unstable bulimic personality-type as its norm, along with the contrasting extremes of obesity and self-starvation. These symbolise . . . the contradictions of the social body—contradictions which make self-management a continual and virtually impossible task. (p. 88)

In high modernity, as Shilling notes, these contradictions appear to be taking the form of a crisis as the body, as a project, begins to confront the limits of malleability. This project of body management is being played out to the point where biological and genetic factors, as well as its ultimate fate, the death of the body, have brought into sharp focus some of the limitations and dangers of the modernist conflation of the body with self-identity (see Fahlberg and Fahlberg, chapter 5). The excesses of the cults of slenderness (Tinning 1985) and appearance (Lasch 1991) discussed by Bordo are hyper-real examples of the playing out of the project of body management, resulting in the increased incidence of eating disorders, over-exercising, and the widespread prevalence of anxiety about body shape (Kissling 1991; Koval 1986; Abraham 1989). These corporeal dimensions of high modernity are, in certain respects, merely a working out of practices constituting and constructing the malleable, modernist body, and of the conflation of the conforming and productive body with social identity, though in high modernity the body and self-identity are increasingly conflated also.

School physical education and sport are in decline, at least in part, because they represent a series of modernist bodily practices concerned with normalising and regulating children's bodies through methods and strategies which are perhaps already culturally obsolete. Australian researchers have shown that a significant number of adolescents are preoccupied with physical activity as recreative sport and entertainment, but at the same time disapprove of their physical education and sport experiences in schools (Tinning and Fitzclarence 1992). This is hardly surprising given the hyper-real representations of sporting bodies through media sport and, in particular, the collages of sporting action which have become a regular feature of mainstream commercial television sports shows (e.g., "Wide World of Sport") and on satellite television beamed into public places. Bradbury (1993) has labeled this phenomenon the "pornography of sport," presenting an objectified performing body which is a seductive counter-example for young people to the mundane routinised bodily practices of school physical education and sport lessons.

However, as Shilling (1993a) argues, perhaps the move beyond exercise, fashion, and diet to more radical surgical interventions for cosmetic purposes has signaled the biological limits of the modernist project of the body. Once the nips, tucks, and sucks have been endured, what other alterations can be made to bodies in order for them to become "normal?" Will some form of genetic manipulation be the next step? Short of this, it seems that cosmetic surgery has exposed the extent to which the body may be malleable and open to human intervention, and has thus brought into sharper focus the interrelatedness of the biological and social. In new times, the hyper-reality of two-dimensional representations of body shape, juxtaposed with the biological delimitations dictated by genetics, have begun to suggest there are indeed limits to the body as an individual, unfinished project.

Reforming Physical Education in High Modernity

In light of this analysis, what can we say about the form and content of physical education programs in new times? How is physical education to contribute to making and remaking the body in high modernity? Will the subject, as Hoffman (1987) has prophesied, be displaced by recreation and sport activities? Will Tinning's (1994) futuristic scenarios become realities, where school students are required to participate in daily fitness sessions run by human movement science specialists and where sport is organised by part-time coaches with only a two-year accreditation degree from technical and further education colleges? Can modernist physical education be reconstructed to make a proactive contribution to the ongoing making and remaking of the body within the hyper-reality of high modernity? I want to conclude this chapter by speculating briefly on the current condition of physical education programs in Australian schools, particularly secondary schools, and what might have to be done to permit these programs to make a more meaningful and critical contribution to popular physical culture in high modernity.

What kind of relationship can modernist physical education practice have to the hyper-reality of much popular physical culture? As I have argued thus far, my view is that the relationship is increasingly dysfunctional. The massive expansion of youth sport in Australia can be counterpoised against growing concern over a high rate of dropout among adolescents across a wide range of sports (Robertson 1988; ASC 1992). In a recent *Quest* article, Robinson (1990) argues that demoralization in physical education classes is a common and widespread feature of school physical education in the U.S.A. Drawing on attribution theory, he has proposed five changes to the ways in which physical education is presented to students: "(*a*) skill mastery rather than comparative performance, (*b*) less traditional and more individualized instruction, . . . (*c*) cooperative learning . . . along with (*d*) proactive teaching behaviours, and (*e*) reattribution training as a means of helping to remedy student demoralization" (p. 36). I suggest that these strategies provide important signposts for new forms of physical education. At the same time, I suggest they are likely to be ineffectual unless there is a parallel realignment of programs and better use of community resources by schools, and also some changes to the ways in which physical educators are educated.

To begin with, if physical education programs are to retain cultural relevance, they must start to both reflect and contribute more directly to popular physical culture. In the secondary school, the staple diet of mainly team sports offered to students en masse, with some limited choice of activities dependent on what schools can offer, requires radical alteration. In secondary school, students should be offered a wide range of activities in which they contract, on an individual basis, to complete a required number of hours of participation, in order to gain sufficient credits in the subject. Some of these activities may be offered by the schools themselves, some may be provided by neighbouring schools, some may be offered by local community agencies that either provide their own facilities or make use of those in the school. Moreover, it should be possible for some students to gain enough credits through participation in activities out of school time, and for others to complete the required hours during a holiday period, depending on the nature of the activity. Activities could be graded for beginners, intermediates, and advanced participants, with a possible requirement that each student learn at least one new activity per school year. Such an arrangement would of course require high level coordinating and planning skills by physical educators, and closer cooperation between teachers in schools and others in the physical culture industry than has perhaps hitherto been the case, along with negotiated agreements on the levels of participation necessary to satisfy the granting of credits.

Such arrangements would force physical educators to look out towards, and embrace, popular physical culture in new ways. But this engagement with what will in some cases undoubtedly be the hyper-reality of popular physical culture, while it may possibly begin to satisfy the interests of most students, should not be a socially neutral one. If physical education is to make a positive and constructive educational contribution to the making and remaking of the body in culture, then this engagement by both teachers and students must be critical of these very practices. Indeed, it is the presence of critique more than any other element which is likely to differentiate a merely recreational program from an educational one (see, Fernández-Balboa, chaper 8). Given this, physical educators will need to be trained to be socially critical of both popular physical culture and the place of their work within it. They will need to be able to see the bigger picture in which students' various experiences of physical activity will be located. As George Sage (1992) has argued, sport scientists and physical educators trained through sport science programs need to "make problematic the dominant discourses, values, and traditions in our everyday lives, by encouraging reflective analysis of and moral deliberation over the dilemmas of sport" (p. 93). He explains:

> Critical social thought applied to sport is not critical simply in the sense of expressing disapproval of contemporary sport forms and practices; instead, its intent is to emphasize that the role of sport scientists needs to be expanded beyond understanding, predicting and controlling to consider the ways in which the social formations of sport can be improved, made more democratic, socially just, and humane. (p. 93)

If this does not happen, the current cognitive dissonance resulting in life-style disorders such as bulimia and excessive exercising and anxiety about body shape are only likely to be perpetuated. It is also likely that physical education practices in schools will continue to be dominated by the discourse of hegemonic masculinity (Sage, chapter 2; von der Lippe, chapter 3; Theberge 1985; Connell 1990).

Physical educators in Australia are becoming aware of the influences of popular physical culture on their students, including the social production of body imagery. In Queensland and Victoria, senior school syllabi in physical education include the critical analysis of the social construction of gendered bodies in and through sport and exercise. The new National Statement and Profile in Health and Physical Education creates the possibility, for the first time, for physical and health educators to cross-reference movement skills with sexuality and other dimensions of health, and to teach towards learning outcomes which problematise compulsory heterosexuality and hegemonic masculinity (Glover 1993). Schools need to take these opportunities to proactively debunk and demythologise various aspects of contemporary popular physical culture which present idealised and glamorised body images. Through cultural critique, they must teach diverse meanings of healthy bodies (Fahlberg and Fahlberg, chapter 5).

In light of the possibility of constructing new configurations of the discourse of embodied femininity and masculinity, it seems to me that there is a need for such reforms which will modify the content, pedagogy, and organisation of school physical education programs. While there has been continuing expansion of the range of available activities which constitute physical education programs in Australia, many continue to be dominated by traditional activities like athletics and major team games. In Victoria, a recent review of physical education has called for a reversion to teaching the "basic" skills of sport participation and for participation in team games to be compulsory for all students (DSE 1994). However, the high visibility of sport and exercise in Australian culture can be an asset in this process of reform. Perhaps, as I have already suggested, a greater degree of choice of activity for students, some of which might be taken up in community sites such as local recreation and exercise clubs, would certainly come closer to matching students' diverse interests in physical activity and increasing disinterest in a diet of team sports (ASC 1991).

At the same time, we need to acknowledge that the association of sport and hegemonic masculinity is extremely powerful and pervades Australian culture (McKay 1991). It would be naive to suggest that physical education programs, which are embedded in the discourse of hegemonic masculinity, might be easily reformed. Nevertheless, there are elements of physical education which are actively challenging hegemonic masculinity, particularly the growing seriousness with which physical educators are approaching the question of health (e.g., Fox 1991). But this work cannot be done by schools alone. These teachers need the assistance of all education, sport, and health professionals to create and reinforce new forms of school physical education which reconfigure the discourse of hegemonic masculinity to empower boys and girls to engage in physical activities in ways which are socially and healthfully enhancing rather than threatening.

If such arrangements at the secondary school level can begin to engage teachers and students in popular physical culture more fully, critically, and constructively, then students in primary schools will need to be prepared to take full advantage of the range of movement opportunities and experiences which will be available to them. Again, the staple of team sports and little choice remains a major obstacle to physical education as it prepares students for a critical appreciation of physical culture. Here, it seems to me that a whole range of physical competencies and a physical literacy of skills and knowledge are central to this preparation. Moreover, if such a scheme is likely to be coherent and workable, physical educators must begin to plan programs which are consistent, systematic, and progressive across the primary/secondary divide. While statewide syllabi that permit such progressions have been in place for some time, institutional structures have worked against their implementation. Physical educators need to be prepared to lobby education authorities to change this situation. Political action is necessary (see von der Lippe, chapter 3).

Only some elements in the reform of physical education for new times have been sketched here. However, if these suggestions seem unrealistic and overly ambitious, then I propose that Hoffman's (1987) and Tinning's (1994) scenarios for the twenty-first century are likely to be right on the mark. In light of the historical positioning of physical education as a site in which the modern body has been made and remade, its continuing educational viability and social relevance are clearly in question if popular physical culture is moving on and young people are moving with it. Some aspects of the hyper-reality of high modernity may well deserve to be resisted by physical educators, such as the mythology of the "trim, taut, and terrific" body. But such resistance and critique will only be effective when they constructively engage with the substance of such myths and practices. A school subject disengaged from the framework of cultural myths and practices can only come to be seen as outmoded and, finally, obsolete.

Closing Remarks

The apparent disjunction between corporeal practices in schools and those in other social sites, such as media sport and exercise gymnasia, points to a lack of uniformity in the development and diffusion of corporeal power in new times, where changes in some sites are more obvious and pronounced than changes in others. Temporal lag and unevenness present particular difficulties for theorising about the body in society. The example presented here of the lag between physical education and sport programs in schools and recent developments in contemporary physical culture suggests the existence of countervailing trends in the operations of corporeal regulation and normalisation. Given Shilling's (1993a) sensible observation that social location, the *habitus*, and the development of taste each play a part in the particular ways in which bodies are socially constructed and constituted, carefully contextualised studies of embodiment in a range of social settings might provide useful perspectives on practices of corporeal regulation and normalisation and the elaboration and diffusion of corporeal power. They may also throw some light on the nature of the transitions currently taking place from modernity to new times across a number of social sites.

The extent to which bodies might be shaped biologically and culturally raises the question of the continuing cultural relevance of school physical education and sport programs. Perhaps only radical alteration to the institution of schooling might permit physical education practices to articulate more coherently with corporeal practices in other sites. Alternatively, school physical education and sport might continue their decline, to be replaced by the kinds of programs foreshadowed by Hoffman and Tinning, which emphasise mere physical fitness and sport performance without a substantial educational core. Or perhaps programs will become privatised, available only to those children who can afford to pay. Given the high commercial stakes of media sport and the increasing numbers of sports scientists in Australian tertiary institutions in place of physical education teachers (Tinning, chapter 7), this commodification of school programs coupled with the commodification of bodies may arrest the decline of physical activity programs in schools. It is open to question whether these new practices will continue to be enmeshed in the new institutions of internal surveillance and the flexible accumulation of high modernist capitalism, and whether the looser form of power they might embody will be individually and socially productive means of corporeal normalisation and regulation.

One final point might be made about the part physical education has played historically in regulating the individual and the social body. As the arguments outlined in this chapter suggest, social regulation of the body has worked to render the body compliant and productive. Foucault's (1977) ideas in this respect are evident in the earlier drilling and exercising form of physical education. With the liberalisation of physical education between the wars, which was crystalised in the Victorian *Grey Book* of 1946, this process of regulation appears to be shifting from external to internal compulsion. In the 1990s, we are now in a position to witness the extent to which this internalised locus of control is diffusing throughout society in, for example, the high levels of body shape anxiety exhibited by secondary school students (Kirk and Tinning 1994). There can be no question that some degree of regulation of the body is inevitable and necessary, but it is essential that such regulation does not become oppressive. Physical education in high modernity has a role to play in both respects, in terms of assisting young people to gain physical competencies which are necessary and important to a full, productive, and happy life, and in terms of alerting them to the potential dangers that exist in any form of internal regulation which might become obsessive or compulsive. Herein lies the challenge ahead for a culturally relevant physical education in new times.

CHAPTER 5

Health, Freedom, and Human Movement
in the Postmodern Era

Larry Fahlberg and Lauri Fahlberg

Introduction: An Emancipatory Interest

The dialogue, speculation, and evidence concerning a possible
transformation from the modern era to a postmodern era has been
considerable. Regardless of where we find ourselves in this transformation—or whether such a transformation is, indeed, in progress—
there are major changes taking place socially and globally. For the
purposes of this chapter, we take the following changes and shifts as
evidence of such a transformation: (*a*) from primarily a technical interest to also an emancipatory interest, (*b*) from primarily technological rationality to also include emancipatory reason, and (*c*)
from body/mind dissociation to bodymind integration. Of course, a
transformation to postmodernism suggests not that these shifts are
totally new, but rather that the proportion of people using, for example, emancipatory reason is now much greater. Our aim is to use
emancipatory interest as a perspective from which to revisit the
dogma of the modern era concerning health and human movement.[1]
Our premises are: (*a*) human freedom is possible; (*b*) human development beyond socially adjusted levels is possible; and (*c*) human
movement, through bodymind integration, can be a facilitator of this
freedom and development.

One of the characteristics of the modern era was the use of technological rationality to the near exclusion of emancipatory reason
(Wilber 1995). By "technological rationality," we mean the rational
process used to determine the most effective means for *any* purpose

65

having a technological interest. By "emancipatory reason" we simply mean the rational process in which emancipation can be realized by bringing critical scrutiny to bear on unquestioned and limiting assumptions, as well as bringing self-reflection to bear on unconscious processes and content. Both technological rationality and emancipatory reason have their benefits. When used alone, however, technological rationality has led to major catastrophe. In a glaring example, technological rationality has been used to discover the means to effectively exterminate large numbers of people deemed unfit to live as in the Holocaust. On the other hand, it is emancipatory reason that provides one of the means to question such a goal and the initial assumptions that provided its foundation. Technological rationality is not problematic; what is highly problematic is using only technological rationality when emancipatory reason is also required. For example, a technical process and a experimental research method can help determine what type of exercise facilitates body fat reduction, but an emancipatory interest can help explicate the social and psychological dynamics that compel many people in our culture to have an unhealthy obsession with exercise as a means of weight control or attaining the "perfect" body—an obsession that limits health and freedom.

A technological imperative in human movement has been the promotion of exercise with an expected outcome of health. While researchers and practitioners typically support the concept of exercise as a "health behavior," the concept of health has seldom been questioned or recognized as value-laden. As in the more glaring example above, the goal (i.e., "health") and its initial assumptions, have remained relatively unquestioned. Subject to emancipatory reason, however, we realize that, far from being an absolute material entity, any concept of health is context dependent. Yet much of the current literature and practice in the area of human movement, as it relates to health, is based on research informed by assumptions and values that have been neither recognized nor subjected to the critical scrutiny of emancipatory reason. Rather, these unquestioned assumptions have been institutionalized, becoming part of the myth of the profession. As Harre (1981) pointed out, "So we are presented with something as if it were empirical, which is heavily loaded with unexamined metaphysical and moral/political presuppositions" (p. 10).

Here, we wish to illuminate these dark halls of myth with both the light of reason and the flame of the revolution in human movement that is now underway (e.g., Johnson 1994; Roth 1989; Bonheim 1992). To accomplish this task, we will first use emancipatory reason to highlight some existing assumptions, particularly as they pertain to the practice (not praxis)[2] of exercise prescription for "health." Second, we will introduce a framework for healthy and free human development, describing some ways in which movement is related to this development.

Within the confines of mainstream modern Western culture, human movement has been advocated predominantly for preventing disease and enhancing sport performance (Brustad, chapter 6; Hellison 1983; Maguire 1991). Such a technological emphasis on a relatively unquestioned (mythic)[3] outcome is illustrative of a profession involved in its practice (not praxis), thus leaving us with a narrow view of movement. In the midst of this emphasis on exercise and sport, the relationships of human movement to human freedom and *full* human development have remained largely unrecognized, unexplored, and therefore elusive, although there has been long-standing dissent (e.g., Bain, Hellison, Lawson). Likewise, the emancipatory potential of human movement has remained relatively untapped. Movement—like many other behaviors—can be recognized as a means of enslaving people and preventing development or of freeing people and facilitating development. For example, movement can facilitate health and happiness, as in the self-transcendent bliss sometimes experienced while running, but movement can also inhibit health, as in a bulimic purge also sometimes associated with running.

Domination related to movement can take many forms. From an emancipatory perspective, perhaps one of the most sacrosanct and widespread health-related examples of domination is the practice of exercise prescription. To lend this prescriptive domain a degree of respectability that it would not otherwise attain, exercise prescription has been propped up by science, and related findings are referred to as "empirical." Manipulating or dominating others to behave in the way we wish by prescribing exercise to them, while ignoring the potential for freedom, is to confuse this domination with a "science" of human behavior and a "health" goal. As such, exercise prescription provides an example of this process of domination so common in the professional practices of the modern era (Lawson 1993a). This process of domination specifically includes mythic premises and goals, technological rationality used to achieve these goals, and experimental and quasi-experimental methodology to demonstrate efficacy. Historically, critical theorists have denounced this prescription-domination relationship:

> One of the basic elements of the relationship between oppressor and oppressed is *prescription*. Every prescription represents the imposition of one man's [*sic*] choice upon another, transforming the

consciousness of the man prescribed to into one that conforms with the prescriber's consciousness. Thus, the behavior of the oppressed is a prescribed behavior, following as it does the guidelines of the oppressor. (Freire 1970, p. 31)

Likewise, one of the original developers of critical theory, Marcuse (1970) wrote:

Domination is in effect whenever the individual's goals and purposes and the means of striving for and attaining them are prescribed to him [*sic*] and performed by him as something prescribed. Domination can be exercised by men, by nature, by things—it can also be internal, exercised by the individual on himself, and appear in the form of autonomy. (pp. 1–2)

For domination to continue, the mythic assumption that health is only a material object that can somehow be conceptualized as value-free has to remain above question and, therefore, beyond the emancipatory light of reason. This assumption then becomes a "given" of reality (myth) and, rather than being recognized as a value-laden concept, the myth of "health" is then accepted as a material object—an absolute—that can then be studied solely through empirical means.

This socially constructed absolute truth, characteristic of modernism, can quickly become the dogma by which oppression occurs (Fahlberg 1995). Like religious zealots, some health professionals believe that they have, and know, absolute truth (after all, it is "scientific") and unquestioningly set about the task of converting "sinners" who are not behaving properly so as to help them achieve an optimal state of "health." The objective is to create new believers in clientele, as well as "true believers" (Hoffer 1951) in professional preparation programs at universities and other training centers. These true believers, armed with the commandments of "health behaviors" and exercise try to manipulate people's actions. Moreover, by using technological rationality they discover ever more clever means to convert these "sinners." Within the belief system that informs the body or related research, the people who are most effectively and consistently manipulated are considered "healthy"; they have complied and they are adhering. With only a technical interest and a mythic outcome, complying to exercise edicts and prescriptions may be inferred or assumed by some to be a natural and positive behavior existing independently of social construction.

In contrast, with an emancipatory interest, freedom takes the place of compliance or adherence. The objective of an emancipatory interest is people who are free, not people who are complying. In fact, "generalization of compliance behavior" (Perkins and Epstein 1988) is anathema to emancipatory interest and is an issue to be problematized. The critical inquiry of emancipatory reason aims to explicate just such hidden power relationships, distortions, and value assumptions.

Culture-wide change appears to be afoot, and, in what may be seen as a postmodern turn, many professionals in the health sciences are now recognizing a broader conceptualization of health that includes distinct human aspects. Once the fullness and richness of this human world is recognized, it can be seen that this world goes beyond mere materialism to include human consciousness, beyond social adjustment to include what Maslow (1969) referred to as "the further reaches of human nature," and beyond both oppressive and repressive forces of imprisonment to include freedom.

Many of us live imprisoned much of the time in a variety of ways. At least two of the jail house bars are cast by psychological and social forces. Typically, we have various degrees of awareness of some of these bars; whereas other bars escape our awareness and, therefore, limit our possibilities to free ourselves. For example, psychologically, we are imprisoned and influenced by unconscious forces, whereas, socially, we operate with unexamined assumptions that limit our freedom (Fromm 1969; McCarthy 1981). But it is not necessary to remain so confined; a new, more exalting vision can include the path leading beyond social oppression and psychological repression. We will define health as the freedom to travel this path, in accord with developmental potential.

Adult Development

Operating within the bounds of cultural constraints and the limiting assumptions of modernism, human growth and development, in general, has also been limited; the Western world has been slow to integrate approaches for optimal development, preferring instead to focus on social adjustment as a single criterion of evaluation. Unfortunately, using social adjustment to assess human development obscures and hinders the potential to grow beyond what is socially reinforced and what sometimes may be the psychopathology of the normal (Fromm 1969; Horney 1942; Maslow 1969). Our new vision, ever more possible in the postmodern era, needs to include a view beyond mere social adjustment and the psychopathology of the normal, especially if "normal" does, indeed, turn out to be sick. As E. Fromm (1955) pointed out, "the fact that millions of people share the same forms of mental pathology does not make these people sane" (p. 23).

Emphasis on movement as a means to enhance freedom (e.g., self-expression or self-transcendence) has been distinctly lacking in both the Western culture and the Western academy and science. Consequently, the use of movement for adult development has been relatively limited. Once we adopt a developmental approach, we can recognize that humans can grow beyond what is commonly accepted in the Western culture as optimal adult development.[4] Due to space limitations, we will simply present examples of some aspects of personal and social freedom in relation to movement. To overcome the limiting assumptions of the modern era, we will emphasize those aspects of movement that are *human* (e.g., freedom, self-actualization, self-transcendence) rather than merely biological or mechanistic forms (e.g., physiology, biomechanics). In other words, our focus will be on the *human* moving, rather than on the *movement* of the human (Rintala 1991).

Once we recognize that humans can freely grow and develop, we can begin to legitimize movement and health as expressions of freedom, rather than focusing on compliance and behavior change (as if we were training dogs to fetch). We may then see that we have been prisoners, sometimes of our own device, and that we have failed to recognize the intrinsic qualities that make us different from, as well as more than, biomachines. Otherwise, we will continue to impede our own growth, health, and freedom.

From Behavior to Consciousness

Central to any discussion of a postmodern approach to health and human movement is the study of, and attention to, human consciousness. Given a human-centered definition of health that includes consciousness and freedom, changes in awareness (rather than only changes in behavior) can be recognized and facilitated. In our view, human consciousness is the medium for growth in a developmental framework that recognizes the potential for the evolution of consciousness through the life span.

Consciousness, at any given time, refers to the immediate and total awareness of who we are and of the nature of the world in which we live. Consciousness is our current, most basic and unquestioned sense of "reality" and our place in that reality. Without self-reflection, we typically lack awareness of the relative nature of our consciousness. That is, without critical self-reflection on our own consciousness, we are not aware that other levels or states of consciousness exist (Ingham, chapter 10; Tart 1987).

Once we recognize human consciousness, a central issue for emancipation becomes the nature and power of the unconscious. In psychodynamic theory, the depth psychology hypothesis holds that a tremendous amount of mental activity happens unconsciously. Although the unconscious plays anywhere from a small to a large role in influencing perception and behavior, we are not consciously aware of this influence. Making both our individual and collective unconscious available is part of the work of the self-reflection of emancipatory interest.

Limits to Freedom: Social Oppression and Psychological Repression

From a developmental perspective, emancipatory interests cannot afford to be focused solely on the external world of social oppression when psychological repression also limits freedom. Critical inquiry (Hellison, chapter 12), for example, has also focused on the internal world of psychological repression (Marcuse 1970). More recently, Wilber (1981) has written:

> It is ridiculous to suggest that, in such extreme cases of exploitation as literal slavery, the victim is secretly responsible. But in so many lesser and more subtle forms of exploitation, the oppressed are indeed "in love with their chains." The Frankfurt school—aided by psychoanalytic insight—spent its early years redressing just that imbalance, and showing that, in many aspects of oppression, the oppressed secure their own chains and hand the key to their future oppressors—"the hidden unconscious tie," said Marcuse, "which binds the oppressed to their oppressors." . . . The point, simply, is that the self is already anxious to repress itself, and since the internalization of oppression helps produce extra repression, the self is *partially* a willing victim from the start. (pp. 268–269)

According to Freire (1970), this " 'fear of freedom' which afflicts the oppressed, a fear which may equally well lead them to desire the role of the oppressor or bind them to the role of oppressed, should be examined" (p. 31). It is just such an examination that provides the balance to critical social theory and provides our rationale for including repression whenever we speak of oppression—the look "without" (oppression) is also a look "within" (repression). As Wilber (1995) has pointed out, consciousness (including critical consciousness) creates as much as it discovers. To clarify what we mean by social oppression and psychological repression, we will briefly describe them below.

Social Oppression

According to Freire (1970), "Any situation in which 'A' objectively exploits 'B' or hinders his [*sic*] pursuit of self-affirmation as a responsible person is one of oppression [because] it interferes with man's ontological and historical vocation to be more fully human" (pp. 40–41). This vocation to be more fully human, through an emancipatory interest and development, is precisely what we are addressing in this chapter. An objective of a sociological emancipatory interest is to make oppressive relationships explicit while recognizing them as socially constructed rather than as given or natural (Allen 1985).

To provide an example of oppression in human movement studies we need only to look at technological rationality, which is particularly oppressive when it is used as if it were a value-free means of generating knowledge. Through emancipatory reason, we can now see that positivistic inquiry has been a way of "concealing a commitment to technological rationality behind a facade of value-freedom" (McCarthy 1981, p. 8; see also Brustad, chapter 6). So we are presented with facts as if they were recently discovered natural laws, rather than as the empirical products of cultural and social construction. Namenwirth (1986) put it this way:

> By draping their scientific activities in claims of neutrality, detachment, and objectivity, scientists augment the perceived importance of their views, absolve themselves of social responsibility for the applications of their work, and leave their (unconscious) minds wide open to political and cultural assumptions. (p. 29)

Once again, we want to point out that the problem is *not* technological rationality itself. Rather, the problem is that it is promoted as a value-free means of inquiry when it is actually value-laden (Bain 1989).

From a critical perspective, the denial of values in human movement results in more distorted "knowledge." Such a distortion needs to be clarified through critical inquiry (see Hellison, chapter 12) and emancipatory reason, lest it results in not only oppressive knowledge but also oppressive practice. As in the case of exercise prescription, the value framework of the professional needs to be made explicit to avoid, among other things, oppressive social expectations arising from knowledge with a technical interest which is being presented as value-free. In our example, people are then admonished to "buy into" this view through adherence to, and compliance with, an exercise prescription that would supposedly enhance their "health." In this process, people escape from their own freedom by allowing the "expert" to make decisions on their behalf (Fromm 1969). Yet, if we do not wish to escape from freedom, we must call upon emancipatory reason to facilitate our questioning of the unrecognized cultural, gender, and socioeconomic biases and assumptions of exercise prescription.

In the above, we can note that oppression was allowed to proceed through at least two fundamental errors of omission. First, the technological rationality informing positivistic and experimental research is exempt from emancipatory reason. Second, the biased knowledge generated is not subject to the critical reflection of *praxis* (Freire 1970) prior to using this knowledge. Because neither of these correctives of emancipatory reason are used, knowledge generation becomes oppressive.

Psychological Repression

As a psychology, an emancipatory interest draws from those psychologies that posit the possibility of freedom from deliberating or maladaptive unconscious influences. To the extent that these unconscious influence result in limitations to freedom, they need to be brought into consciousness so that their limitations can be recognized and assuaged. Although unconscious forces typically go unrecognized precisely because they are unconscious, they play an important role in psychodynamic and other emancipatory perspectives that recognize human consciousness and the unconscious (e.g., Berger and Mackenzie 1980; Fahlberg, Fahlberg, and Gates 1992).

Like any activity, human movement can be used for repression—to blot out an awareness of the deeper, disturbing issues of life. Typically, repression is no longer maintained when the forward momentum of activity is no longer possible. As one female exerciser reported:

> I got injured in a motorcycle accident and it was really hard. I was burned out too, and I got sick and had a virus. But I just never quite got it back together, so I ended up in the hospital with this virus, and, boy, it seems like my life just fell apart. Couldn't exercise, couldn't do anything, just lethargic and it was like my life had come

to an end. . . . It was almost like a blackout. It's like, I can't really remember a lot of it, other than that I was depressed. And I wanted to get out. It was almost like a fog. You know, kind of feeling like I don't know what's going on. . . . (Fahlberg and Fahlberg 1994, p. 106)

Exercise, as a means of repression, works as long as one is able to exercise. When exercise is no longer possible—typically due to injury or illness—a crisis can ensue.

Freedom and Consciousness in Evolution

The developmental framework that we describe below focuses on the evolution of consciousness (Fahlberg and Fahlberg 1991; Grof 1988; Pelletier 1985; Singhe 1988; Walsh 1980; Wilber 1975, 1977, 1981, 1995). For the possibility of freedom for all, as well as for optimal individual and collective growth, the maps describing the evolution of consciousness represent a significantly different way of thinking about health and human freedom, both individually and collectively.

In the evolution of consciousness, there are a number of identities (Wilber 1975). These identities can be represented by where we implicitly draw the boundaries of who we are at any given stage of development (Wilber 1979). Early stages, for example, are characterized by narrow boundaries and a sense of separation, while later stages are characterized by expansive boundaries and a decreasing sense of separation from others, the environment, the planet, and the universe (Wilber 1979, 1995). Because these states of identity are strongly represented in consciousness as who or what we are, they are referred to as levels of consciousness or developmental identities.

Within our total awareness or consciousness there is a core sense of who and what we are in relation to everything else (Wilber 1981). When we conceptualize who and what we are to ourselves and others, we are expressing our "self concept," usually in terms of age, gender, ethnicity, occupation, religion, nationality, and so forth. These reference points and contexts make up our basic identity. In other words, the "normal" adult ego in our culture consists of all of these reference concepts and relationships by which we identify who and what we are, usually in some form of a physically encapsulated "person," separate from everything else (Wilber 1975).

Human development, from birth to death, can be viewed as a vast spectrum of consciousness stages in relation to these identities (Wilber 1975, 1977). While Western theorists mapped initial stages of human development, Eastern observers (at least three thousand years ago) began mapping stages of human consciousness evolution during the second half of life (Wilber 1977). With the newly found postmodern ability to synthesize across cultures, we can now combine these two perspectives emphasizing that which involves development beyond the adult ego stage (Wilber 1977).

Once again, it is imperative to recognize the central and profound revolutionary notion that humans, given appropriate conditions, can grow beyond the adult ego (Fahlberg and Fahlberg 1991; Fahlberg, Wolfer, and Fahlberg 1992). Although theorists have mapped many precise and specific stages of consciousness development, for the purposes of this chapter, we will outline only four identity stages which we will refer to as (*a*) persona, (*b*) ego, (*c*) existential, and (*d*) transpersonal. Rather than describing these stages in detail, our brief characterization focuses on those dimensions which are more pertinent for highlighting the possible relationships among freedom, health, and human movement.

Persona Identity

The "persona" identity (Jung 1971) represents an early stage of adult development. This persona is a "good face" or a "social mask" which a person uses to facilitate social interaction. In other words, the "persona" is the facade personality that is merely a response to social demand. Neumann (1969) explained:

> The persona, the mask, what one passes for and what one appears to be, in contrast to one's real individual nature, corresponds to one's adaptation to the requirements of the age, of one's personal environment, and of the community. The persona is the cloak and the shell, the armour and the uniform, behind which and within which the individual conceals himself [*sic*]—from himself, often enough, as well as from the world. (pp. 37–38)

As a result, the persona is fraudulent—an inaccurate sense of self which is created when we are in denial of our tendencies (e.g., anger, joy, hostility, erotic impulses, courage, aggression, assertiveness, drive). When these tendencies are denied they do not disappear; rather, they remain present in the unconscious. However, by repressing these tendencies, it may *appear* to us not to have these unwanted tendencies, resulting in a partial sense of identity ("I") or the persona. The unacceptable "dark side" is still present, but it is repressed. C. G. Jung (1971) referred to this dark side as the "shadow."

The "shadow" is composed of unconscious contents that are deemed unacceptable in whatever circumstances one lives. When the shadow and persona are not integrated, the ego views itself as only a "shining light" which lacks a shadow. Therefore, the shadow is the opposite of what we view as ideal. We confuse ourselves with a facade personality that has been formed according to social acceptability, and we "forget" our shadow elements (Neumann 1969).

"Projection" occurs when we fail to recognize alienated traits or tendencies as our own. Although these traits or tendencies are ours, we perceive them as belonging to someone else. The aspects of the self considered "bad" are split off unconsciously and projected onto others in an effort to rid the self of identification with these aspects (Ogden 1982). Our sense of self then becomes extremely narrow, and the unwanted tendencies are excluded. Once the shadow has been denied and subsequently projected, we identify with what is left— the persona.

According to Wilber (1979), there are *two major consequences* of this shadow projection. First, people feel that they totally lack the projected tendency, and second, the projected tendency is seen to exist as a negative characteristic in others. These tendencies that bother us or attract us in some way, are often our shadows projected onto others. Despite our efforts, however, we cannot erase these tendencies; we can only deny "ownership" of them (Wilber 1979).

The *resistance* to integrate our shadow and claim ownership of its unwanted contents can be very strong. Wherever there is projection, there is also resistance, which can range from mild to severe (Wilber 1979). To return to an earlier example, we may deem reasonable the extermination of an entire ethnic group based on our projection upon them of our shadow material. This ethnic group, or "evil empire," then embodies all of the "evil" and receives all of the blame for the "wrongs" of the day. But it is important to note that the blame which we lay at their feet would be the projection of our own tendencies. In colloquial terms, "the pot would be calling the kettle black."

If professionals, for example, are stuck in the persona, efforts may be directed toward changing other people (e.g., clients, students, spouse, relatives) and their behavior, even though these efforts are based on authority rather than reason. For example, the professional in human movement might focus on the necessity for participants in a cardiac rehabilitation program to "take responsibility for their health" by increasingly being more compliant with an exercise prescription, never noticing the contradiction: being compliant with an authority is not taking responsibility. With no attention to self-transformation, including consciousness and the repressed unconscious, we have no opportunity to experience enhanced freedom through self-reflection (i.e., reflecting our projects back to ourselves). Our shadows remain "other."

Often, the shadow can be seen by what is observed in the world and the reaction that ensues. As in the case of critique,

> Critical judgment is not necessarily shadow projection. But when the finger is pointed, it is useful to look not only at where it points, but also back at the finger pointer to see what motivation and benefit might reside there. Self-elevation by denouncing others is so tempting and gratifying, and so universal, that no condemnation of apparent evil should be taken entirely at face value. (Yandell, in Neumann 1969, p. 4)

Self-reflection is crucial for those adopting a critical perspective (Ingham, chapter 10). The pivotal questions are: What is the difference between projection of the shadow and critical social theory? How do we know that the social injustice we see "out there" is not also something that we ourselves perpetuate? This is precisely the point at which the aspect of critical reflection in praxis, born of emancipatory interest, also includes critical self-reflection (Held 1980; Horney 1942; Jung 1971; McCarthy 1981; Neumann 1969).

If we can accept the role of the unconscious, we may then engage in critical self-reflection to question the nature of the "evil" that we see, and whether or not it is *only* "out there." To achieve individual and collective freedom, it is essential to recognize the need to integrate the persona and the shadow, thereby accepting even the socially unacceptable aspects of our being. Continued development is contingent on bringing maladaptive unconscious material into consciousness so that it loses its ability to negatively influence behavior and limit freedom. The process of human growth and development, then, is a process of expanding our sense of self, sense of identity, and conscious awareness, through combining, at this stage, the persona and the shadow to form a strong ego.

To integrate the persona and the shadow we must "own" our projections (Gordon 1968). But tapping into the unconscious mind directly does not work for many people (Hannah 1991) because appeals to the mind to integrate the persona and the shadow are appeals to that same entity that alienated them in the first place. In contrast, using movement to become aware of, and integrate, the shadow can facilitate the circumvention of mental censorship. We can use movement to integrate the shadow in a number of ways. As an example, movement therapy is often used for human growth and development. Since overt activity is often linked to repressed content, transformative movement experiences can provide direct experiential access to those aspects of the self which are not otherwise readily available, such as the shadow. Through an expansion and intensification of movement experiences, repressed content can be encountered and integrated by moving the body in different ways from those to which we are accustomed (Kleinman 1978; Sansom 1972; Schmais and White 1968; Siegel 1973), but we need to be aware that by doing so we are "stirring the pot" of both repression and oppression. When we move consciously and mindfully, "anger, despair, fear, and sexuality may rise to the surface and ask to be accepted and integrated" (Bonheim 1992, p. 94).

It is important to note that transformative movement works because the body and the mind are integrated. That is, while in the modern era body and mind were not only differentiated but dissociated, the postmodern era provides a space for bodymind integration.[5]

For transformative purposes, it is important to recognize that there is no right or wrong way to move. According to May (1974), "we learn that technique can be used as an intellectualizing defense against understanding of the self" (p. xi). Hence, emphasis on performance or skill can be absolutely detrimental to transformation (May 1974). One possible way to address this issue "is to affirm to yourself: 'I am moving for myself, for my own pleasure. This is a gift to me!' When you understand this, there is no pressure to do it right to please the instructor or even your own inner critic" (Bonheim 1992, p. 26). Put another way, when there is no right or wrong way to move, and no "good" or "bad" person as a result, there is no persona to be fed by social expectation. There is no mask to don. The veil of repression may then lift, and whatever has been repressed can become conscious. The shadow comes into the light, thereby ceasing to be the shadow, and, as its contents are recognized as one's own, they can be integrated to form the ego.

Ego Identity

From an egoic perspective, personal worth is based on a conceptualization of who or what one is (Wittine 1989), as typically defined in our culture by values emphasizing individuality (Bellah 1985);

wealth, power, and prestige (Walsh and Shapiro 1983); competition, self-assertion, effectiveness, and productivity (Grof 1988); as well as social roles and community position. Our identification with these various characteristics, along with the social norms which establish personal and social hierarchies, give us a sense of identity in terms of who we *think* we are and the role we play in our melodrama of self-worth. In this melodrama, existence is reduced to what we *have* and *do*. The egoic level of consciousness is a vulnerable identity with attachment to pleasure and avoidance of pain as its basic dynamic. If an idea or person threatens our ego, our usual reaction is to defend and attack (Wilber 1981). Anything that challenges the conceptual categories of who we are is a threat to our existence and reality.

The prevailing conviction in Western societies is that, once we reach the level of the well-adjusted, effective, and "productive" ego, the end of development has been reached and that no further growth is possible. In contrast, from an expanded, evolutionary perspective, achieving a well-balanced ego is just another developmental stage, and we have the potential to grow beyond this stage.

Western psychology holds that strong ego development is necessary for optimal health. Just so, typical movement strategies (e.g., competition, performance enhancement) are often oriented toward strengthening the ego; human movement professionals generally value and applaud methods that enhance ego-based "self" constructs (e.g., self-esteem and self-concept). When the ego is considered the end point of development, it is much more difficult to recognize those aspects of movement that are both oppressive and repressive. In contrast, a developmental perspective allows us to recognize the ego as just another stage of development. When we recognize the possibility of further development, we can clearly see oppression and repression merge to form a nexus of limitations to health and freedom.

As we pointed out above, the basic dynamic of the ego is attachment to pleasure and avoidance of pain. The most painful awareness of the ego is its own demise or death. Yet, the ego can overcome this awareness in a variety of creative ways (Becker 1973), which we will refer to as "immortality projects" (Wilber 1983). Fear of death creates the deep and powerful need for immortality projects in the form of both secular and religious activities. Examples of immortality projects and how they can be both oppressive and repressive become evident in at least two themes related to human movement. First, exercise becomes an immortality project when it is prescribed for the prevention of death. Efforts to prevent death become evident in many of the professional journals and publications on human movement, as it relates to exercise prescription for health. For instance, the American College of Sports Medicine (1991) claims that, without exercise, "many more lives would be lost due to the deleterious effects of sedentary living" (p. 3). It is important to note here that death is being denied as an inevitable part of life; these professionals are actually attempting to *prevent* death! A crucially important point is that the crusade to prescribe exercise to "save lives," without acknowledging the inevitability of death, influences behavior unconsciously. With an emancipatory interest, this denial is not to be indulged; instead, it is to be brought into conscious awareness so that we can be free of the enslaving unconscious influence that such a denial perpetuates.

The second theme—repression of the awareness of death—is also primarily unconscious. At the egoic level, we keep the full realization of death at bay through strong and intense identification with, and involvement in, our personal activities and projects. We engage in socially useful and rewarded projects (e.g., jobs, professions, service, productions, contributions, etc.) and tranquilize ourselves with the trivial (Kierkegaard 1849). The trivial can take a variety of forms including substances (Peele 1985) (e.g., food, drugs) and experiences (e.g., mass entertainment, work, busyness, exercise, etc.). If we can keep "busy" through movement and other activities, the inevitable (i.e., death) can be successfully repressed. However, the fear of death begins to emerge when "the forward momentum of activity is no longer possible" (Becker 1973, p. 23). As Yates, Leehey, and Shisslak (1983) point out:

> Obligatory runners and anorexic women [*sic*] must continue to prove themselves by running or dieting. . . . They are satisfied by moving toward a goal, not by achieving it; in fact the goal itself is entirely secondary and is reset at will to rationalize the continuation of the process. . . . The endless quest . . . is perpetuated by the fear that if one stops, one will cease to exist. (p. 225)

In this case, the forward momentum toward a goal helps to repress the ever-present but unconscious fear of death. This repression occurs psychosocially in that it is difficult for an individual to become truly aware of death in a culture that denies it so strongly (Pelletier 1985).

Transformation from egoic to existential consciousness represents a profound shift entailing a growing recognition and awareness that it is impossible to fully live until we are also aware that we will die. As Becker (1973) contended, we cannot possibly know how to live until the reality of death percolates to the core of our being. If the reality of death is denied, then life is distorted. Once we accept death as a reality and as a valuable part of life, further growth to the existential stage is possible.

Existential Identity

The group in the West most responsible for bringing emancipatory interest to bear on the issue of death has been the existentialists. At least two major and overlapping themes can be found in the literature concerning the existential level of development. The first theme is the recovery and integration of our concrete, organic sense of self (body) with our socially learned conceptual self (mind); a process involving the conscious realization of our immortality. The developmental theme here is bodymind integration, not merely bodymind interaction. Remember that we were stuck in the egoic stage to the extent that our conscious awareness consists largely of conceptual, abstract categories (e.g., professor, student, athlete) rationally organized as egos, and also we experienced our "essence" through these abstractions. Armed with these abstractions, our body and the natural world were experienced as separate objects: We were our minds; but we owned our bodies (Wilber 1979). Now at the existential stage, the *realization* (as opposed to mere *intellectualization*) of death comes with great discomfort and anxiety. During the existential stage, a central "developmental task" is to reunite our abstract, conceptual self (mind) with our concrete, organic sense of self (body). As this integration continues, the body (and "its" physical world) is less feared, less rejected, and no longer needs strict control. To the extent that this integration occurs, we are liberated from a mere conceptual existence to a full-bodied existence. We "real-ize" that we are much more than that which we merely *thought* we were. In the process of integrating our bodymind, we also experience directly the unavoidable fact of our eventual physical death. Although realizing that death as an integral part of life creates existential suffering, it also enables a fuller engagement with the miracle of existence in the here-and-now.

The second theme concerns the growing recognition of the arbitrariness and contingency of our socially determined and maintained conceptual self. While the mental egoic stage was characterized by learning to adapt and adjust to society (May 1983), the existential stage is characterized by the transcension of social conditioning in

82 Larry Fahlberg and Lauri Fahlberg

the process of finding the genuine self. Recognizing our social conditioning is the first step in the process of freeing ourselves from this conditioning and developing an autonomous and authentic self. That is, moving from egoic to existential identity is, in part, a process of removing ourselves from socially-constructed limitations and reinforcements, freeing ourselves to choose our possibilities, and taking responsibility for creating our own meaning and value (Fahlberg, Fahlberg, and Gates 1992; Washburn 1990).

Existential growth involves increasing authenticity and freedom, combined with a decreasing need to conform and belong. Becker (1973), referring to S. Kierkegaard, wrote: "Kierkegaard had no easy idea of what 'health' was, but he knew what it was not: it was not normal adjustment—anything but that, as he has taken such excruciating analytical pains to show us. To be a 'normal cultural man' is, for Kierkegaard, to be sick . . ." (p. 86). By definition, then, people emphasizing authenticity are less likely to support domination simply on the basis of social conformity or irrational appeals to "health." That is, mythical claims of health as a commodity are subject to critical scrutiny. At the ego stage, all of these difficult issues are avoided through simply doing what one is told or going along with the crowd (Fromm 1969); thus securing one's oppressive chains. In this regard, Gruba-McCallister (1991) explains, "By allowing ourselves to be defined by others' opinions and standards, we avoid the terror involved in becoming a self" (p. 80). At the existential stage, there is no escape from this terror, no matter how many miles we run. Likewise, there is no escape from the freedom and responsibility of creating our meaning in life.

Authenticity implies freedom within the bounds of given constraints and is, therefore, not fulfilled or reinforced through group membership or the conformity to the social norms and ideologies that such membership requires. The emphasis on authenticity allows the existential identity to minimize the needs to belong—needs that can be so readily met through automaton conformity. As May (1983) suggests, "To the extent that my sense of existence is authentic, it is precisely not what others have told me I should be . . ." (p. 102). Unlike socially conditioned behavior, the authentic being is constituted by intrinsic meaning. People at the existential level realize that they are worthy simply because they exist, rather than because they produce or conform to whatever it is that society happens to value at the time. For example, when we engage in movement for the sake of movement, not to achieve a socially dictated goal, its intrinsic value emerges.

When using maps of consciousness, differentiations are typically made according to where the boundaries of awareness lie. In the map we are using here, the existential identity is the final stage of *personal* development; beyond this stage, development becomes *transpersonal*.[6] That is, once the existential level of identity is transcended, the transpersonal self begins to emerge.

Transpersonal Identity

Transpersonal theorists suggest that it is possible to evolve beyond the separate, isolated self to a deep sense of being as an integral part of, or "one" with, all life. The term "transpersonal" means beyond the personal sense of self. The transpersonal "self" is a center of awareness ". . . which is not exclusively identified with one's mind or body or ego . . ." (Wilber 1983, p. 102). A major developmental aspect of the transpersonal stage is the transcendence of the separate self and, hence, the relief from egocentrism and ethnocentrism, as well as from existential anxiety and despair. When the transpersonal self begins to emerge, we are able to observe emotions and thoughts without strongly identifying with them.

When this transpersonal mode of being becomes our primary mode of consciousness, our ego and existential identities become integrated into transpersonal consciousness. To be sure, our ego and existential selves are still functional, but we no longer identify primarily with those aspects of our being. In other words, an expanded sense of developmental integration is realized which includes but transcends earlier, less-expansive integrations. For example, with transpersonal consciousness there is no longer a primary identification with a separate ego—an ego with psychological and materialistic demands to enhance and protect itself and with psychological and materialistic demands and strategies for self-aggrandizement, defense, and attack. The sense of "us *vs.* them" dissolves into "we." We lose the deep psychosocial needs that lead to repression and oppression, and link our inner transformation to outer perception and expression.

Movement can enhance this transformation through what are now being recognized as movement-induced peak experiences.[7] Although there is a vast panoply of practices that facilitate peak experiences (Wilber 1993, 1995), our focus is on movement as a way to cultivate both peak experiences and structural adaptation to transpersonal identity. That is, peak experiences are temporary experiences of transcendence as well as expressions of a more permanent and stable adaptation to, and acquisition of, transpersonal identity (Wilber 1993, 1995). For example, movement as a trigger for peak experiences is evident in a variety of forms including hatha yoga (e.g., samadhi) and running (e.g., the runner's high). As Bannister (1955) wrote, "I sometimes think that running has given me a glimpse of the greatest freedom a man [sic] can ever know, because it results in the simultaneous liberation of both body and mind . . ." (p. 265).

Peak experiences can be conceptualized as temporary insights— what Wilber (1983) refers to as "peek" experiences into the transpersonal: Blissful epiphanies that are often also numinous, noetic, and illuminative, enhancing insight and understanding. This bliss may be relatively easy to attain through movement meditation (e.g., Bonheim 1992), and may have a tendency to happen as much during movement as in other activities (Berger and McInman 1993). The fullness of this movement experience can be profound and life-enhancing (Berger and McInman 1993; Park 1973, McInman and Grove 1991) through a sense of unity with the planet and a profound sense of connectedness with others (Howze, Smith, and DiGilio 1989).

However blissful and insightful these peak experiences might be, transpersonal identity represents transformational growth, and stable, structural adaptation to the transpersonal, not just transitory (albeit profound) experience (Wilber 1983). So we have a peak experience, which can occur in virtually anyone, and we have structural transformation and adaptation to transpersonal identity, which appears to be quite rare. Nonetheless, as Wilber (1995) points out, "temporary peak experiences from higher domains . . . can indeed *accelerate transformation* toward those domains" (p. 750). The difference between having peak experiences, on the one hand, and achieving the stage of transpersonal identity, on the other hand, is the process of converting fleeting peak experiences into an enduring structure (Wilber 1995).

As in other stages of development, there are obstacles to freedom, both oppressive and repressive, that await the traveler on this road. When movement as a transpersonal practice becomes rigidified, it also becomes oppressive. That is, the same authoritarianism of a teacher at the ego level can also bedevil a teacher whose espoused aim is to facilitate transcendence to the transpersonal. For example, due to the nature of transcendent experiences, these are sometimes interpreted in an authoritarian, patriarchal, religious context and according to religious dogma, rather than being interpreted in a spiritual context that is highly idiosyncratic and democratic (Bonheim 1992). In the former, one may be exhorted to perform certain movements in specified ways, lest movement-induced transcendence elude one forever (Johnson 1994). This type of admonition or prescription represents the same domineering oppression that is observable at the persona and ego levels, exemplified by the exercise prescription. Furthermore, when we are oppressed we may be also helping to secure our own chains, thus maintaining our personal repression by submitting to these edicts. In such cases, consciously or unconsciously, we escape freedom and the unknown void that freedom may represent (Loy 1992). Like words, the variety of movements in which we engage can become mere activities that distract and defend us from the threats we sense in the unknown, including unknown freedom. However, movement can also expand our reality beyond words. This is the reality of the transcendent and the identity of the transpersonal.

All of these identities have corollaries in how freedom is conceptualized and experienced. In the early stages of development (i.e., persona, ego), freedom might be experienced as the ability to make personal choices and act on them, to respond to our desires (often at the expense of others). With an emphasis on *only* these conceptualizations of freedom, we think that "taking what we want, getting what we want, or doing what we want is freedom" (Fahlberg 1995). In contrast, from an existential perspective, we may view socially conditioned personal desires as deeply internalized traps to be transcended through authentic, albeit painful, growth. Finally, at the transpersonal stage, we may experience freedom as transcendence of the bodymind. With the transcendence of identification with concepts (e.g., freedom), freedom itself becomes a realization rather than only an intellectualization or passing experience. Freedom becomes who one *is* rather than something one *gets*. The body and the mind are recognized as only temporary phenomena, waves on the ocean of consciousness that will eventually calm and die.

Closing Remarks

As dualisms that plagued the ear of modernism begin to fade, we may more easily recognize that the reflection in praxis is always also self-reflection. In a society in which the persona may be the pacer of development (Wilber 1981), it is crucial that the critical finger of praxis—pointing out social justice—also inspires conscious self-reflection to prevent projecting our "unacceptable" characteristics "out there." To escape the bars of repression and oppression, and to make freedom an experience and, ultimately, a realization, we need both critical self-reflection and critical reflection, informed by emancipatory reason, in our personal, professional, and social lives (Tarnas 1991).

To recognize the need to break free from psychosocial constraints, including the constraints that limited world views and paradigms place on our collective and individual developmental potential, we recognize freedom as a value, a means, and an outcome for health and human movement. From an emancipatory perspective, the emphasis in regard to human movement is *not* on getting people to behave—to adhere or to comply—in the manner in which we dictate. Rather, the emphasis is on facilitating freedom and enhancing the conditions by which this freedom may be maximized. Because this view represents a radical departure from *modern* orientations to movement, the following points deserve careful emphasis: (*a*) both human freedom and development beyond socially acceptable levels are possible; and (*b*) we can use emancipatory interest and human movement to facilitate both this freedom and this development.

If our definition of "health" includes the possibility of enhanced freedom and the evolution of consciousness, then it seems that the human movement profession would benefit from a recognition of these human possibilities. Our personal, educational, and research activities could then support our individual *and* collective freedom. This is an important point, for we are *not* trying to create another dogma or convert people to our world view. Rather, we consider the map of consciousness that we have described as a useful tool for freedom and developmentally appropriate conceptions of health and human movement. Likewise, it is important to emphasize, particularly in our individualistic culture, that as our individual and collective growth are intimately linked, so, too, are our individual and collective freedom. Our focus in this chapter has been on the individual and collective, personal and transpersonal, psychological and social. As R. Walsh and F. Vaughan (1993) put it, "The aim is to illuminate the destructive psychological and social forces that have brought us to this turning point in history and transform them into constructive forces for our collective survival, well-being, and awakening" (p. 231).

CHAPTER 6

A Critical-Postmodern Perspective
on Knowledge Development
in Human Movement

Robert Brustad

Introduction

In all disciplines, the generation of new knowledge is essential
to the growth and professional advancement of the field. In comparison to other academic disciplines, systematic research and inquiry
in human movement are relatively recent phenomena and present a
unique challenge due to the diverse physical, psychological, and social processes that influence learning and behavior in movement settings. From a critical perspective, one must also consider the means
and meanings of human movement, not to mention the influence of
economic and political dimensions of social life as they may impact
opportunities for physical activity across groups differing by race,
gender, socioeconomic status, physical ability, and so forth.

Such considerations suggest multidimensional research approaches and compel us to question the value of narrow or simplistic
explanations. Of particular benefit here is the postmodern perspective, which emphasizes sociohistorical and cultural analyses and the
need for integrative, inclusive, and dynamic approaches to knowledge. However, as a consequence of traditional research training and
a failure to closely consider key epistemological assumptions and
issues regarding the unique character of our field of study, most
human movement researchers have relied upon conventional (i.e.,
positivistic) research approaches. This has hindered the development of a broad knowledge base and a full understanding of professional practice.

This chapter will examine the current status of knowledge development within our profession focusing on three issues. First, how we study human movement. I will give particular attention to traditional assumptions from the positivist scientific model and will identify some of the limitations of this approach. Second, where we study human movement. Here I will argue that, too often, the real social world of human movement has been ignored; I will discuss the lack of "context-rich" research, and I will suggest some possible contexts for inquiry. Third, what we study in our field. What we study obviously reflects what we perceive to be important—that is, the development of skilled movement performance, at the expense of the developmental, affective, and social processes and outcomes that accompany human movement.

How We Study: Research Paradigms and Human Movement Research

The nature and influence of scientific paradigms upon knowledge generation has been a hot academic topic since the publication of Thomas Kuhn's (1962) *The Structure of Scientific Revolutions.* Kuhn's use of the term "paradigms" referred to basic world views, or core belief systems, about how the world operates and, from a scientific standpoint, how knowledge about the world should be gained. Most of the scientific inquiry in the twentieth century, has been based on the positivist paradigm and its dominant model of "orthodox" science (Martens 1987). The assumptions and scientific processes of the positivist model are most evident in the research methods of the biological and physical sciences.

Although no single, universally accepted definition of positivism exists, the most striking elements of this world view are the belief that reality is best understood through the use of systematic, experimental methodologies (Dewar and Horn 1992; Hoshmand 1989; Sparkes 1992). A key doctrine of positivism is objectivity. Objectivity is based on the concept that it is possible to gain knowledge about the world by maintaining a detached, emotion- and value-free posture regarding that which one is studying (Sparkes 1992). Reductionism is another core concept of this paradigm. Reductionism is the attempt to understand the functioning of the whole through an analysis of its individual parts. As such, the reductionistic approach provides a mechanistic model for understanding both the natural world and human behavior. In essence, positivist scientific methods clearly reflect the modernist world view that a singular, universal, unchanging reality exists independent of sociohistorical context and individual perspective.

Traditionally, human movement researchers have embraced the methods and assumptions of the positivist model (McKay, Gore, and Kirk 1990). This is attributable to several factors. First, to the successes of the positivist paradigm in addressing many research problems in the natural sciences. Second, to the strong foothold that the natural sciences have in human movement studies (Harris 1983a; Harris 1983b). Unfortunately, many researchers hold the "naive belief that the epistemologies and methodologies of natural science can be transported unproblematically into social and cultural settings" (McKay, Gore, and Kirk 1990, p. 55). Third, to the need by human movement studies to gain recognition and acceptance in the scientific community.

Yet, the positivist paradigm is both limiting and limited. It greatly structures the types of questions that are posed by human movement researchers—questions that are generally devised in ways that reflect an experimental intervention (e.g., treatment and control groups, concern for internal validity, etc.). In this vein, hypotheses are generated in advance, and quantitative forms of data which are amenable to tests of statistical significance are employed. Moreover, positivistic inquiry is incapable of explaining human movement behavior in relation to social, cultural, and contextual forces and has a limited capacity for critical reflection. Hence, in order to study these phenomena, alternative research paradigms are needed.

Although such paradigms do exist for the study of human movement, they have received limited attention in our profession. Students in departments of physical education and sport studies receive minimal or no exposure to alternative qualitative approaches of scientific inquiry (Bain 1989). Such approaches include interpretive, critical, and transformative inquiry. These alternative qualitative models focus upon understanding the meaning of human behavior with reference to social and historical contexts (Bain 1989; Sparkes 1992) and stem from the sociological traditions of symbolic interactionism and ethnomethodology, relying typically upon ethnographic and naturalistic methods for understanding human social behavior. A core component of the qualitative paradigm is the idea that reality is socially constructed (Sparkes 1992). Thus, because they presume that knowledge generation is a social process and is context-specific, qualitative researchers do not seek absolute answers about the nature of reality. What they do seek is to describe "reality" from the point of

view of individuals as they experience it. Examples of qualitative research in human movement will be discussed in the following section.

Within the qualitative paradigm, the interpretive method stops
at merely describing events and interactions, whereas research carried out from critical and transformative theoretical perspectives
directs its primary attention toward understanding the nature of
power relationships that exist within a society. Hence, critical and
transformative researchers are not merely interested in understanding social relations but also are committed to changing and improving social structures. Whereas the mode of inquiry may be the same
as in interpretive research, the goal of the critical and transformative research is to go beyond description and to use knowledge as a
means of enhancing the quality of movement experiences for individuals, and bringing about equity and justice.

As such, positivistic science itself (as the only conception of
"good science") is the object of critique and deconstruction by critical-
postmodern theorists and scientists. They argue that the dominance
of the positivist perspective yields a number of problems regarding
the value of knowledge. Major concerns involve the principles of reductionism, objectivity, and quantification. Another concern relates
to an imbalanced reliance upon deductive methods at the expense of
inductive approaches.

Moreover, critical researchers in our field argue that human
movement must be studied in relation to social and political contextual arrangements and ideologies, adding that to adopt a narrow,
microscopic view is to limit understanding of these core influences.
The positivist approach results in the research spotlight being trained
at the individual level of analysis, which inherently encourages psychological explanations of behavior while minimizing the role of
social forces. This individual focus cannot account for human interaction nor for social circumstances.

The tendency to focus on the individual, while disregarding the
broader social network of influences, represents a major trend in all
forms of social science research over recent decades in the United
States (Gergen 1987; Sampson 1977). This occurrence is not surprising because it is consistent with a growing cultural ideology of "self-
contained individualism" (Sampson 1977, p. 769). The benefits of a
more "contextualized" knowledge base will be addressed in a following section.

Another major critique of the positivist paradigm deals with the
notion of objectivity. The idea that researchers can, and do, study
phenomena objectively, and independently of their own social, his-

torical, and political frame of reference has been widely challenged (e.g., Bain 1990a; Krane 1994; Kuhn 1962).

> On the positivist view, therefore, good scientists are detached observers and manipulators of nature who follow strict methodological rules, which enable them to separate themselves from the special values, interests and emotions generated by their class, race, sex, or unique situation. (Jaggar 1983, p. 356. Cited in Dewar and Horn 1992)

In this regard, Martens (1987) maintains that the illusion of objectivity hinders the advancement of social science research because it encourages researchers to work within artificial and sterile settings rather than engaging in "real world" forms of study. Furthermore, such "objective" research lacks external validity because its findings cannot necessarily be generalized to non-contrived settings.

Yet another aspect of the positivist paradigm which needs to be examined is the principle of quantification. Many researchers trained within a positivist perspective sustain that that which cannot be counted or measured is not worth studying. Such a mindset naturally excludes the examination of social, political, and cultural forms of influence. Fortunately, the idea that qualitative data can provide valid and useful information is gaining acceptance (e.g., Locke 1989; Sage 1989). However, qualitative research perspectives, such as interpretive and critical inquiry, are in turn, criticized for subjectivity and experimental bias.

There is one more flaw associated with "positivism in action," and it pertains to the relative use of inductive and deductive methods. Although the "scientific method" is supposed to rely upon both techniques, the reality is that modern scientific research almost completely ignores inductive or interpretive processes (Hoshmand 1989). Typically, research hypotheses are clearly identified in advance and tested accordingly, but rarely do positivist scientists conduct research with observational, interpretive, or reflective goals. Thus, only infrequently does knowledge generation and theory development in human movement studies emerge from observation and reflection about the real world movement experiences of individuals.

It is not my intention in this chapter to discredit positivism as a means of scientific inquiry, but only to highlight the problems and limitations of the positivist perspective as the *single* legitimate means of gaining knowledge about the dynamic social world of human movement. The influence of this scientific view, however, cannot be underestimated. In Martens' words "orthodox science exerts a comprehensive power over most of us today, not unlike religion once did" (1987, p. 36). From a critical perspective, we must recognize the dominance of this scientific world view upon our attempts to understand human actions.

Where We Study: Knowledge Development in Relation To Culture and Context

The framework within which human movement occurs has largely been ignored by human movement researchers. Certainly, little attention has been given to *where* we conduct our research because of our tacit acceptance of the positivist paradigm and its concerns for experimental control. Consequently, we have a limited understanding of the means by which cultural values and situational influences impact human movement behavior. Movement participation does not occur in a cultural vacuum (Krane 1994). In fact, without reference to culture, we cannot understand how individuals differing by gender, age, and race, experience sport, play, and exercise. In addition to broader cultural influences, the immediate social situation, or context, must be examined if one desires to understand the purpose and meaning of movement practices.

The relationship between the cultural context and human movement is evident when examining the participatory patterns of males and females in sport and exercise. Participatory involvement is clearly influenced by Western cultural beliefs about the "gender appropriateness" of various forms of physical activity for boys and girls and men and women. These cultural values are communicated by parents, peers, and other social members and the media. For example, in the United States, parents value physical ability for boys more highly than for girls, providing more encouragement for their sons than for their daughters (Brustad 1993; Eccles and Harold 1991). Although actual differences between boys and girls in physical ability are minimal during childhood (Thomas and French 1985), the general public perceives that there are substantial differences between young boys and girls regarding physical ability and motivation to be physically active (Brustad 1993; Eccles and Harold 1991).

Children's peer groups also exert a strong influence upon children's physical activity involvement. Physical ability is much more highly valued within the peer groups of boys than it is within the peer groups of girls (Chase and Dummer 1992). Because peer groups are also very instrumental in gender role socialization (Adler, Kless, and Adler 1992; Eder and Parker 1987), group values regarding sport and exercise involvement are potent influences upon willingness levels of youngsters to participate in various forms of physical activity.

The physical activity behavior of adult males and females is similarly related to gender-linked cultural influences. For example, research indicates that appearance and weight loss are the primary motives for physical activity involvement for a substantial proportion of American women, but the same does not hold for men (Davis and Cowles 1991). Clearly, in the last decades, changing cultural images of physical attractiveness for women have had a strong impact upon the meaning, type, and extent of the involvement in physical practices by American women.

Although the study of culturally based gender roles in movement practices can help explain the participatory interest and behavior of individuals, it has been quite neglected by investigators in our field. The preponderance of research in this area has been conducted by sociologists and developmental psychologists. One possible explanation for this is that, because of traditional research training, human movement researchers feel poorly prepared to study the socio-cultural influences on physical activity and, therefore, avoid examining some of the major forms of influence affecting movement practices in our society. In addition to cultural values pertaining to gender, we can add a variety of other culturally based influences including the role of the media, perspectives on work and play, and age- and gender-related stereotypes.

The social context is also an important, but neglected, dimension of study. To date, most research in physical education and the sport sciences has been "decontextualized" in that aspects of the social ecology of participation are rarely examined. Components of social context include group characteristics such as size and composition, norms, values, status characteristics, social evaluation opportunities, expectations, and the like. Leadership and teaching behaviors also certainly shape the social context of human movement, particularly as these behaviors influence the psychological, social, political, and affective climate of the setting. Analyses of the "hidden curriculum" in physical education (Bain 1976, 1985, 1990b; Fernández-Balboa 1993a; Kirk 1992a), for example, have provided considerable insight into contextual processes affecting learning and behavior.

The lack of attention devoted to contextual characteristics represents a major critique, as well (e.g., Foon 1987; Gergen 1987; Gill

1992; Levine, Resnick, and Higgins 1993). With regard to the sport sciences, Gill (1992) commented that "our research and practice seems narrower and more oblivious to social context and process than ever before" (p. 155). Furthermore, the need for a close consideration of contextual influences in helping us understand human movement behavior is reflected by Gergen's (1987) statement that "just as individual words cannot be understood out of a linguistic context, the understanding of individuals requires comprehension of social context" (p. 63).

Another important reason to take the social context into account is that the meaning of movement practices can only be understood in relation to the context within which they occur. Physical activity is purposeful and personally meaningful. A great deal of research in this area, however, has been "meaning-less" in that it has frequently failed to consider the personal meaning (Fahlberg, Fahlberg, and Gates 1992) and the interactional characteristics of movement.

A relatively small body of knowledge addressing culture, context, and meaning currently exists. Representative research has been conducted in the areas such as adult exercise involvement (Bain 1985a; Bain, Wilson, and Chaikind 1989), gender differences in participatory patterns in middle school physical education (Griffin 1984, 1985), and Little League baseball players' interpretations of their experiences in relation to the meaning of their involvement in this sport (Harris 1983b). I will discuss the work by Bain and colleagues as a means of highlighting the role of the social context in exercise.

Bain (1985a) studied the effect of the social atmosphere of a university fitness and weight control class upon students' subsequent participation and motivation levels. A major finding from this study pertained to the effect of the class climate on students' levels of participation and their continued interest in physical activity. Through observations and interviews, Bain concluded that the instructional climate of this class could best be characterized as "technical-rational" (Fahlberg and Fahlberg, chapter 5) because the instructor assumed that knowledge about the health benefits of exercise would contribute to greater levels of exercise involvement by students. However, the students revealed that the subjective and affective dimensions of exercise were the more central aspects of their participation.

In a subsequent study, Bain, Wilson, and Chaikind (1989) examined situational and cultural factors impacting the experiences of overweight women in an exercise program. The researchers focused their examination on the role of socially-constructed meanings about fitness, physical appearance, and health as they might affect these women's perspectives on exercise involvement. Employing a variety of data collection techniques (i.e., interviews, group discussions, and journals), the investigators attempted to identify common patterns affecting participation. A principal theme related to body size. The women reported experiencing considerable social disapproval in movement settings based on their size which contributed to heightened negative self-concepts and inhibited and restricted their involvement in many physical activities. These experiences of social disapproval detrimentally impacted the participants' perspectives on exercise and their willingness to participate in physical activity in the future.

These examples make us aware of the need to examine the framework within which human movement occurs. Moreover, it is also evident that in order to understand cultural and contextual forces, we must venture beyond the laboratory and examine movement in natural settings.

What We Study: Performance Enhancement Versus Social Relevance

The third issue to be discussed here relates to the focus of our research. The question, simply put, is: "What is important for us to know?" Given the breadth of our field, there are certainly a great number of possible avenues for study relating to participation, learning, and human development through movement. What we choose to study reflects both current values and research traditions. Although there exists a wide range of potential topics for investigation, human movement researchers have devoted a disproportionate amount of attention to a rather narrow area; namely, the study of high level performance of physical skills (Whitson and MacIntosh 1990). While we know a great deal about how to enhance the performance of elite athletes, we know relatively little about other important topics such as the effects of physical education and sport upon human development (Maguire 1991). On this issue, R. Tinning (1993a) asks:

> Why, for example, are the problems of sports performance dominant and the problems of ethics in sport marginalised? Why are questions of the biological functioning of the body dominant and those related to the social theory of the body ignored? (p. 14)

Whitson and McIntosh (1990) trace the scientific interest in high-level performance to the relatively recent emergence of the sport science model of research. Implicit within this approach is a reliance upon the reductionistic imagery of the "body as machine." Conversely, there has been little concern for the social, political, developmental, and ethical outcomes of sport involvement. Whitson and McIntosh (1990) have described the carryover of this performance-oriented approach to physical education as the "scientization of physical education," in which the research methods, values, and orientations of sport scientists have shaped the form of knowledge and the practices generated in the profession of human movement.

The issue of what we study can also be examined in relation to the concept of "discourse." A "discourse" represents the way reality is portrayed through language and images, and in relation to research, the "taken-for-granted" perspective on what is most important to our field (Tinning 1993a, 1993b). Tinning (chapter 7) describes and contrasts two discourses within our field, those of performance and participation:

> The language of the *performance* discourses is about selection, training, thresholds, work loads, progressive overload, etc., implicitly supporting competition, survival of the fittest, and the exclusion of the less fit or able. . . . The language of the *participation* discourses is about inclusion, equity, involvement, enjoyment, social justice, caring, cooperation, movement, etc. (pp. 97–98) (emphasis added)

Many essential questions pertaining to the interrelationships between human movement and society have also received minimal attention. For example, socially constructed messages about the body, fitness, and health are integrally related to beliefs about the meaning and perceived value of physical activity within our culture and, subsequently tied to the extent and type of physical activity involvement that people pursue. Only a handful of researchers in human movement (e.g., Fahlberg and Fahlberg, chapter 5; Fahlberg, Fahlberg, and Gates 1992) have examined the origin of these socially constructed messages of health and fitness and their effects upon movement practices. For example, of the many dimensions comprising health (e.g., mental, spiritual, emotional), why have physical health indices (e.g., cholesterol, body composition) become the primary criteria for health status in North American society? Furthermore, why have we failed to pursue, or even actively consider, negative outcomes that result from current cultural ideals of health and fitness such as disordered eating, social physique anxiety and identity issues, and the detrimental effects of excessive exercise?

In pursuing socially relevant research issues, there exists the possibility that the knowledge we gain can help us to transform and enhance the quality and level of involvement in movement for a greater number of individuals. For example, we might try to understand how sport participation is limited by racial and gender role stereotypes, authoritarian traditions, homophobic attitudes, and the like; thereby generating strategies for counteracting those forces which stifle participation. Similarly, we may choose to more actively understand the role of movement as it influences human development across the life span (Maguire 1991).

Closing Remarks

The development of knowledge in human movement studies is not value-free. Knowledge development is clearly affected by prevailing research paradigms and scientific traditions, and is tied to particular interest groups. If our goal is to enhance the quality of movement experiences for all, then we must examine more closely the processes by which we generate new knowledge. I suggest three ways of contributing to this knowledge base.

First, that we utilize a wider variety of research paradigms and methods in our study of human movement. In this chapter, I have discussed some limitations of positivist research. The major difficulty associated with positivism has less to do with its methods than with its customary acceptance as the only means by which to gain legitimate knowledge in our field. A postmodernist perspective on research—one that accepts and integrates different ways of knowing—may help us better understand sport, exercise, and physical activity practices.

Second, to be consistent with a postmodern view of multiple representations and interpretations of reality, we can also begin to better understand some of the unique social and personal meanings that influence the quality of movement experiences. Hence, I suggest that we closely examine the purposes and goals of our knowledge development. The sole generation of new knowledge in human movement is no guarantee that we will enhance the quality of movement experiences for individuals. Only if our intent is to critically examine and transform attitudes and practices may we serve and improve society. In this vein, we can ask ourselves further questions about how, what, and where we can accomplish this goal in human movement research. A more democratic and inclusive orientation to our research should bring more beneficial movement experiences to more members of our society.

And third, and also consistent with the postmodern perspective, I urge us to recognize the role that situated reality plays on human movement. We must seek to understand the meanings and purposes of movement in various contexts and also inquire about how factors such as gender, race, and socioeconomic status, physical ability, and so forth, impact movement opportunities and contexts.

CHAPTER 7

Performance and Participation Discourses
in Human Movement:
Toward a Socially Critical Physical Education

Richard Tinning

Introductory Comments and Stories

The term "physical education" is out of fashion as a generic term
for our field. In universities all around the Western world, many former departments or faculties of physical education have changed
their names. In the United States, the favoured name is Kinesiology,
in Australia it is Human movement/sport science, in the United King-
dom it is sports-exercise science, and the University of Granada is
probably setting the trend in Spain with its Faculty of the Sciences of
Physical Activity and Sport. Faculty funding, research moneys, and
other preferential treatments are more forthcoming for faculties that
have eschewed the title physical education, or at least positioned it
marginally to human movement or sports/exercise science.

In this chapter, I will argue that professional practice in human
movement is limited by the discourses which underpin training
courses for future professionals and its conception of what needs to be
done in the name of professional practice. I will pay particular attention to the professional practice of physical education teaching given
its social service role (Lawson 1993a).[1] First, I will present three
short stories which illustrate various ideas contained in this chapter:

The First Story

Jane Clifton, writing in a Melbourne newspaper (*The Age*, 17/4/
95), was lamenting the closure of the old municipal swimming pools

throughout that city. They were being closed, she argued, because they didn't fit the exercise image of the 1990s. They didn't have the necessary gyms, saunas, spas, aerobics room. They needed updating. The local pool "must become a centre for training, hard work . . . a miniature sports institute in every suburb" (p. 20). In addition, "The amicable, rotund, sun-scarred pool managers of old are secretly being replaced by goose-stepping, cardio-funk androids in lime green bicycle shorts" (p. 20).

The Second Story

Andrew Roberts is a recent graduate from one of Australia's leading universities. He has a degree in sports science and a qualification for teaching. He is a keen triathlete and spends many hours each day training to improve his performance. He wears lime-green bicycle shorts. Last year, in the national titles, Andrew placed sixteenth. Since his graduation, Andrew has worked as a physical education teacher in a large secondary school. He finds the job frustrating because "most of the kids just aren't motivated to sweat. They simply don't want to exert themselves enough to improve their fitness." Andrew finds that most of the knowledge he attained in his sports science degree is of little use with kids who are not interested in improving their performance.

Recently Andrew has been looking for another job. Perhaps, if he can get in the top ten in the nationals, he might attract some sponsorship money and be able to devote all his time to training. In the meantime, he thinks that working in a local fitness center might be more rewarding than teaching school physical education.

The Third Story

Meme McDonald (1992) wrote a book titled *Put Your Whole Self In*. It is the story of Meme's encounter, and subsequent involvement, with a self-help hydrotherapy and massage group in the Melbourne suburb of Northcote. The group of women (aged from 60 to 90) participated in various exercises in the water under the direction of a woman named Marj. Marj was seventy-nine and had no degree in human movement or physical education. She did have years of experience in swimming but no lime-green bicycle shorts. The story gives a moving account of how, for these women, the hydrotherapy had transformed their lives. Their arthritic and aching bodies found relief in the support of the water. They also found support in the caring and love of one another.

The women in the Northcote group didn't benefit from any of our new found knowledge about human movement or sports science. They never had their body fat estimated, their oxygen uptake measured, their mobility tested. They didn't need cardio-funk music to work by, nor did they need any systematic recording of their progress displayed on a chart or a computer screen. But they *knew* how they had improved and they *knew* the value of the exercise classes to their total well being. They didn't need the help of a "modern" human movement professional. Indeed, it's doubtful if they would have been any better off if they had been instructed by such a professional.

These three stories contain important issues and trends related to our professional work in the field of human movement. First, Jane Clifton's lament over the changing nature of local swimming pools reveals the trend towards training and fitness improvement as the chief legitimising reason for many human movement related recreative activities. Second, and in sympathy with the first, Andrew Roberts' story exemplifies a growing problem for teachers of physical education who are trained in the science of human movement. Training and fitness has become the raison d'être for their professional work and often this meets resistance and opposition from a majority of secondary school adolescents. Third, the Northcote hydrotherapy group shows that, for countless individuals, improvement in performance is not the reason why they participate in physical activity. Moreover, Meme McDonald's story raises important questions regarding whether all movement professionals in such contexts need to be trained in sports science.

These three stories are in some way connected with our field. They are connected, albeit somewhat abstractly, by the *discourses which constitute them*. The concept of "discourse" is a very powerful and useful one to understand. In this chapter, I am using discourse to mean a recurring pattern of language (or visual images) about a phenomenon which portrays reality in a certain way. It represents a world view that becomes part of the taken-for-granted way in which a phenomenon, such as sport or physical activity, is to be understood (Sage 1993).

Importantly, in all professional practice, some discourses become dominant while others are marginalised or backgrounded. In Western medicine, for example, it is the discourses of science which are dominant while other ways of thinking about the body and health are defined as "alternative" (see chapter 5). The same is true in the field of human movement. Knowing about how discourses work is necessary to understand why our profession has traditionally focused its collective attention on some issues and not others, and why our courses for the training of future professionals are dominated by certain discourses.

Performance and Participation Discourses in the Field of Human Movement

It is useful to think of two major types of discourses which determine our work as professionals in the field of human movement. They are the discourses of performance (after Whitson and MacIntosh 1990), and the discourses of participation. At the outset, I recognise the dangers in attempting to classify (or name) such orientations, for there are always exceptions and confounding, complicating factors to consider. However, notwithstanding the recognition of the limitations of this modernist categorising on either/or, dichotomous bases, I consider that this framework can provide a useful heuristic for thinking about our professional work and the appropriate forms of professional training.

Performance Discourses

Human movement professionals who work as exercise/sport scientists and elite sports coaches are predominantly concerned with improving human performance, most often in relation to sports. The discourses that underpin most of their work are those of science. Science is used as the method (or means) of obtaining improved performance (the goal or the end). The main consideration with such performance-oriented discourses is how can performance be improved or enhanced. Thus, questions of means are dominant. In our universities, these discourses are legitimated through such subjects as biomechanics, exercise physiology, sports psychology, tests and measurements, sports medicine, fitness training, and so on. The language of performance discourses is about selection, training, thresholds, work loads, progressive overload, and so forth, implicitly supporting competition, survival of the fittest, and the exclusion of the less fit or able.

Participation Discourses

I am using the label "participation discourses" to refer to the discourses which underpin the focus or orientation of physical education teachers in schools and recreation workers with the aged, the disabled, or other "special" populations. Their professional mission is to increase participation in the movement culture (Crum 1993) for all the therapeutic and educational values which can be derived from such participation. Improving performance is not their raison d'être. Rather, issues related to improving performance have relevance mainly in terms of how such improvement can enhance participation tendencies and well-being. Accordingly, the knowledge which they draw on most frequently in their professional practice will be that derived from the social sciences (sociology, psychology, social-psychology, anthropology, etc.) and education (teaching, learning). The language of the participation discourses is about inclusion, equity, involvement, enjoyment, social justice, caring, cooperation, movement, and so forth.

The Dominance of Performance Discourses

The performance and participation discourses do not share equal status within our field. Rather, the field of human movement has become increasingly dominated by the science-based, performance-oriented discourses (Burstad, chapter 6). If we look at the "typical" degree program in human movement in most Western countries, we will find that the subjects which are considered necessary for performance-oriented professional roles are dominant (see Macdonald 1992). Because these programs privilege the discourses of science, they reinforce a particular view of human movement. They reinforce a particular view of the world. However, although dominant, this world view is not a universally accurate or accepted view. Indeed, this scientised view has come under increasing criticism in recent years from a number of fronts. There are critiques from within science itself, from social scientists, from philosophers of science, and from scholars in human movement (e.g., Kirk 1990; Schempp 1993; see also chapter 5 in this book). According to Lawson:

> Originally steeped in a broader social morality during the initial stages of the progressive era, today our field is seemingly adrift in a quasi-scientific and technical sea without the wherewithal for morally-informed navigation. (Lawson 1993a, p. 17)

In the same vein, Sage (1993) argues that the world doesn't finish at the outfield fence, the gymnasium walls, or the laboratory door, and that the human movement profession should "consider its role as more than socially reproductive: there is a need for resistance and opposition as well" (p. 162).

Both Lawson and Sage raise important questions about the relationship between the human movement profession and the society which it is supposed to serve. They argue that we must see our professional work as part of the wider social and world scenes, and that we must consider how our work makes for a better world. The discourses underpinning Lawson's and Sage's arguments, and those of this book, are concerned with democracy, shared responsibility, social change, inclusivity, and equity and justice. They are the discourses of participation rather than performance.[2]

Why Do I Argue for a Socially Critical Physical Education?

Chomsky (1989) makes a point with respect to public discourse in politics and sport that, when driving, he sometimes turns on the radio and finds himself listening to a discussion about sports. He realizes that the people who call the radio station during the program do not defer to sports experts. Instead, they have their own opinions and speak with confidence, with a high degree of thought and analysis, about all sorts of sophisticated details about coaches, players, plays, and so forth. Chomsky suggests that, on the other hand, when he hears people talk about international affairs or domestic problems, for instance, the conversations are characterized by a high degree of shallowness. What puzzles him is that he does not think that international or domestic affairs are much more complicated than sports, given that the former type of affairs does not require extraordinary skill or understanding to uncover the deceptive illusions that obfuscate contemporary reality. In fact, he comments, it requires normal levels of skepticism and analytic skills. Paradoxically, people tend to exercise such type of skepticism in analyzing what a certain team or coach ought to do next Sunday instead of in questions that are far more important.

Like Chomsky, I do believe that sports are a relatively trivial issue compared to the civil war in the former Yugoslavia, mass starvation in countries like Sudan, the plight of the poor and homeless in our cities, racial bigotry, domestic violence, rape, drug abuse, or the spread of Aids. Moreover, I have no doubt that sport is a very useful vehicle for deflecting national attention away from issues of the state and political processes. I am not saying that the purpose of sport or physical education is to correct the massive social and economic problems of the world. Obviously that is ridiculous. However, both sport and school physical education are sites of cultural practice within the field of human movement which have the capacity not only to reproduce, but also to challenge the dominant ideologies which underpin violence, poverty, and oppression.

I also support Alexander et al.'s (1993) claim that the purpose of physical education should be to "increase a person's approach tendencies and abilities to participate in a successful, rewarding and socially responsible way, in the movement culture" (p. 17). Unfortunately, the requirement of social responsibility which this definition requires is too often ignored. In my view, human movement professionals have a responsibility to try to identify the ways in which our professional practice affects, and is affected by, social issues such as violence, sexism, racism, or other forms of injustice, and that, with such an identification, we have a moral responsibility to attempt to change our practice in socially responsible ways.

Physical educators (i.e., teachers in school and teacher educators in universities) often argue for physical education for *all* as if it were a kind of universal "good" and consequently beneficial for everybody. They ignore the significant fact that physical education, as it is currently implemented, may be detrimental for many. Like it or not, it often involves problematic practices. Physical education can liberate and oppress, inspire and disillusion, encourage and alienate, and be a source of satisfaction and achievement as well as of disappointment and failure. We as professionals are part of the problem and part of the solution. In a slightly broader context, Fahlberg and Fahlberg (chapter 5) point out that movement itself can be enslaving and prevent development, but it can also be freeing and facilitate development. To understand some of these problematic practices it is useful to analyze the roots of the professional identity of human movement as a field (Sage, chapter 2). Such roots are to be found in the beginnings of the physical education profession in the late nineteenth century (Kirk, chapter 4).

The Roots of Human Movement Professionals

Lawson (1993a) recently delivered a powerful critique of many of the assumptions which underpin professional practice in the human service professions such as human movement. In "The Dudley A. Sargent Memorial Lecture" to the audience at the 1993 NAPEHE conference, Lawson provided a thought provoking analysis of the origins of physical education as a serving profession in the U.S.A., and in particular the part that Dudley A. Sargent played in this history. In his analysis, Lawson claimed that Sargent and many of his contemporaries in the late nineteenth century believed that "ordinary people were inherently weak and feeble, needing to be protected from their own folly and rashness" (p. 3). These early human movement professionals were worried (and rightly so) about the ill health caused by industrialization, and, with evangelist zeal, they advocated "exercise programs aimed at restoring and maintaining the bodily health of the masses" (p. 3). Lawson went on to say that "Sargent, like so many others in our field, believed that without compulsion and regulation, persons needing these [exercises] the most would not experience them, [and that] without professional regulation, the health, lifestyles, and lives of ordinary people would be adversely affected" (p. 4). The discourses championed by Sargent were characteristically modernist in that they reflected a grand narrative (science as a panacea), and they privileged the expert in terms of valued knowledge. Kirk (see chapter 4) makes a similar argument with regard to the regulation and compulsion of the body in Australia.

In the early twentieth century, physical education was one of the so-called human service professions along with nursing and social work. It was a time in which modernist thinking dominated conceptions of professional practice. Lawson (1993a) puts it this way:

> At the same time that Sargent was offering health-enhancing exercises to people presumed to be in need of them, other human service professions extended, in parallel fashion, mental health, education, family preservation, appropriate nutrition, health education, "worthy" use of leisure time and the like to these same people. . . . The root metaphor was the machine and the assembly line: For each specialized part of the human being and human life, at least one specialized profession emerged to both define and serve needs, problems and wants. (p. 5)

Moreover, Lawson claims that, at that time, the human service professions operated under what he terms a "human capital model" that was informed by conceptions of humans as another form of capital in the context of markets. In such a conception, the profession attempts to capitalise on a market of human "need." The essence of Lawson's argument is that, at least in the early twentieth century, the human service professions like human movement set out to *regulate* the lives of people in their own best interests. There was a mix of paternalism, evangelicalism, and arrogance at work here. It was experts (professionals) who "knew" what was best for people and they alone could provide the right cure (Fahlberg and Fahlberg, chapter 5).

Favoured Questions, Favoured Knowledge

We can learn a lot about a professional field by studying the sorts of questions it asks and the problems it considers important. It is instructive to examine which problems are considered by the field of human movement to be important and which are marginalised. Why, for example, are the problems of sports performance dominant and the problems of ethics in sport marginalised? Why are questions of the biological functioning of the body dominant and those related to the social theory of the body ignored? Why is it that in research on physical education teaching the "problem" of academic learning time has received more attention than issues relating to social justice and equity? Why are performance discourses dominant and participation discourses marginalised in training courses for human movement professionals?

The Question of the Body

One of the most obvious examples of the knowledge bias of our field is in the study of the human body. Given the significance ascribed to the body in postmodern culture and the central place of the body in human movement, it seems axiomatic that future human movement professionals, whatever their specialisms, also understand and appreciate the social, political, and economic nature of their subject matter. Yet, professionals in human movement have centered the study of the body almost exclusively within discourses of science. We study anatomy, physiology, sports medicine, and biomechanics, and these subjects are all based on a particular conception or way of theorizing the body. These are functionalist and foundationalist views of the body (Shilling 1993a). The body is considered an object to be dissected, measured, probed, maintained, tuned, and compared. Our basic understandings of the body are embedded in the analogy of the body as a machine, an analogy that is reductionistic and mechanistic.

In a special issue of *Quest* (1991), readers were introduced to the growing literature on the sociology of the body which challenges

the orthodoxy. This literature is based in social science and underpinned by discourses that differ from the biomedical understandings. In it, scholars in social and feminist theory inform us that
bodies are not *merely* biological organisms but are inscribed with
social meaning—bodies are socially constructed (Shilling 1993a,
1993b), gendered (Dewar 1987; Theberge 1991; Connell 1987), and
are embedded in *both* nature and culture (Kirk 1993, see also chapter 4 in this book).[3]

The Question of "Training"

The value of knowledge informed by the performance discourses
is linked in a circular fashion to the structure of all university
courses. In many universities, scholars who value the performance
discourses have leadership positions and, accordingly, have a greater
influence on the structure of their programs. Students who pass
through these programs learn to value certain knowledge perspectives over others and, in turn, when they graduate and take on professional responsibilities themselves, they tend to perpetuate these
same perspectives.

If we look critically at professional training programs in human
movement, we can learn a lot about the types of knowledge that our
field considers most essential for practice. Also, the analysis of our
professional training courses tells us about how we define or set
problems, what we consider to be legitimate, and which problems
can be solved through the application of our professional skills and
knowledge. On this point, Lawson suggests that our field's leaders
"appear to have erred in their problem-setting [and that such an er-
ror] is not without consequences in today's world" (1993a, p. 18).

Consider this within a human movement program in a typical
higher education institution. A university course which trains future
professionals is actually a response to particular problem-setting—
What is the necessary knowledge that a professional needs in order
to practice her/his profession? This question implies a choice between different views of what knowledge is essential for professional
practice. As such, problem-setting (Lawson 1993b, 1984) involves a
form of social editing where some themes are eliminated from consideration and other themes are foregrounded and become the focus
of attention (see von der Lippe, chapter 3). Problem-setting is, then,
a political act intimately linked with power, control, and what counts
as legitimate knowledge in the culture or profession.

In many contemporary training programs of kinesiology or exercise science the education of teachers has become relatively marginalised. Teaching, as a professional activity, is considered less important than scientific inquiry. Even the most worthwhile (read essential) subject matter content for teaching physical education is centered on the subdisciplines of human movement, in particular those which focus on performance and exercise science (Tinning 1995). Hence, it is not a coincidence that so many physical education teachers now focus their attention on fitness and sport-skill training in their lessons (Peiró and Devís 1993). In other words, the problems which they attempt to solve through the application of their professional knowledge are predominantly those relating to fitness development and sports-skill development.

Macdonald's (1992) study of a human movement degree program in a large Australian university is informative here. Macdonald found that most faculty and students considered that science-based subjects (those based on the performance discourses) were more important for their professional training than subjects in the social sciences (those based on participation discourses). This perspective relating to worthwhile knowledge was manifest in such things as the status accorded faculty members (e.g., research grants, expensive labs, promotion), the amount of curriculum time devoted to different knowledge fields, and the effort students put into studying particular subjects. The "important subjects" (science-related subjects like exercise physiology) received more study time and were treated more seriously by students.

Similar findings were made in Swan's (1993) study of another major PETE department in an Australian university. He documented that some 60 percent of the faculty of the Human Movement department were science-oriented and only 12 percent social-science-oriented. The department's journal holdings were 70 percent in science and these accounted for some 78 percent of the department's library budget. In addition to equipment purchased from research grant moneys obtained via competitive grant applications over the last decade, the department had spent over $480,000 on high-tech equipment for the exercise sciences. In contrast, the social sciences had received almost nothing. When students were asked which subjects in their degree they considered to be of most value to their future careers, over 90 percent listed the subjects of descriptive and functional anatomy, physiology, and exercise physiology. Not one student chose the social science grouping! One student even claimed that "Physiol [*sic*] is the basis for teaching physical education" (Swan 1993, p. 19).

But the way in which science is privileged within the profession is more subtle and ideological than the obvious presentation of the number of courses human movement students take in their training or the dollar value of equipment purchased or grants obtained in such departments. For example, I wonder what sense would Andrew Roberts make of the stories I described in the beginning of this paper? With Andrew's view of the world of human movement, would he scratch his head over the concerns of Jane Clifton? Would he support the "upgrading" of the local pools which Clifton laments? I guess so. Also, Andrew probably would not be particularly interested in Meme's story of the hydrotherapy class because working with the aged (especially with aging/ed bodies) would not reinforce his preoccupation with youth, training, and fitness. Moreover, Andrew would probably consider that Marj should be suitably trained in order to teach and that the members of her class would benefit a lot more from having movement science applied to their swimming.

Reasonably, the way in which professional knowledge is characterised and presented to teachers in training is part of the reason that Andrew Roberts (and those in his position and with his background) finds it so hard to adjust to the professional expectations of contemporary school teaching. As Schön (1983) pointed out, professions which privilege technocratic rationality are finding that their graduates are ill-prepared for many of the tasks that professional life asks of them. Their way of perceiving and understanding physical education is through the lenses of sports performance. Moreover, the sort of professional knowledge that would help them work in a service-oriented environment has been so marginalised in relation to the sciences within their professional training that graduates are ill-prepared for their chosen professional practice. They are left with the modernist discourses of science and its underpinning ideology of technocratic rationality to help them engage the complex postmodern social world in which their professional work is located. Andrew, with his particular performance-oriented lenses, framed the major professional problem in his teaching by blaming it on his students' inability to value hard work and the development of fitness. He never asked the question "could it be that the orientation of the program is inconsistent with the interests of the majority of the students?"

Changing Times, Changing Contexts

In the late twentieth century, a period variously known as post-modernity (Lyotard 1984), late modernity (Giddens 1991), or high modernity (Kirk, chapter 4), Sargent's ideas are called into question. As such, Sargent's ideas relating to the value of physical exercise in the lives of ordinary people, while not explicitly concerned with pro-ducing a docile body,[4] were nonetheless implicated in the capitalist interests of the State.

According to Kirk (1993), Foucault has argued that many of the institutional forms of control and discipline which characterized life in the early 1900s have been replaced by individualized forms. Al-though the State still regulates our lives in various ways, much of the surveillance previously done by it in the early twentieth century is now done by the individual. We now act as our own police in cer-tain matters of control and discipline. The care of the body is one such example. Some of us "choose" to control and discipline our bod-ies through diet and exercise regimes because we have been influ-enced by the discourses relating to the prevention of heart disease and/or physical attractiveness.[5] In this regard, Fahlberg and Fahl-berg (chapter 5) provocatively question whether exercise is an au-thentic mode of self-expression or a response to social conditioning. Kirk (chapter 4) has provided evidence of the former. Although the State regulates our lives with respect to the wearing of seat belts and non-smoking in government buildings, as yet there is no arm of the State which compels us to have slim mesomorph bodies.

However, the governmental policy continues to attempt to regu-late lives through physical education programs in public schools. Kirk and Spiller's (1993) social history of the development of physi-cal education the schools of the Australian state of Victoria from the beginning of this century to the 1950s is informative in this context (see also Kirk, chapter 4). They show clearly how there were various influences on the exercise curriculum (read physical education) which, in most explicit ways, were intent on regulating the lives of school children for "their own best interests." Likewise, Kirk (chap-ter 4) expresses:

> The emergence of various systems of rational gymnastics towards
> the end of the eighteenth century . . . and their eventual widespread
> adoption by a number of institutions such as schools and the military
> by the end of the nineteenth century, was a constituent part of the

development of a range of regulative and normative practices aimed at schooling the docile body. (p. 42)

In recognising that any coherent functioning society requires some regulation, we can easily understand how schools play a vital role in the construction of a regulated life. Certainly, most societies tell their young what is good for them and regulate their lives accordingly. But what happens when people leave the institutional controls of school? How does the human movement profession continue its influence through social and individual forms of regulated life in the postmodern era? More and more, we are seeing a shift of influence from schools to more general cultural forms, particularly mass culture. The rise of mass culture (especially through the electronic media) is a feature of postmodernity, and the fact that school physical education exists simultaneously within both mass culture and education becomes a very significant contextual consideration.

Among the main missions of the field of human movement are to educate people in the care of their bodies and to help them engage in "healthy" life-style pursuits. Some of this education takes place via physical education classes in the public institutions of schooling. But as Kirk and Spiller (1993) have shown, schools not only educate, they regulate—they regulate the body through, among other things, physical education. Regulating the body is, however, a less significant function of contemporary schooling (and physical education) because nowadays much of the work in regulating life beyond the institution of compulsory schooling is done by the media and alternative organizations (e.g., "health" clubs) with vested interests in the body and or physical activity.[6]

Physical education teacher education is now replete with calls for a more socially critical curriculum for the preparation of future teachers (Fernández-Balboa, chapter 8; Pettit 1992; Bain 1990a; Tinning 1987; Kirk 1986a, 1986b), particularly in relation to issues of equity and justice. Yet, this remains largely rhetorical considering that the majority of physical education teachers, teacher educators, and others working in the field of human movement do not consider social justice and institutional change as parts of their professional mission.

But even if the specific courses in teacher education (such as curriculum and pedagogy) themselves became more socially critical, their impact would be marginalised in the context of the general faculty's reverence of the performance discourses of exercise/sport science. Accordingly, changes in teacher education must be accompanied by changes in the way the field of human movement initiates its future professionals into the knowledge of the field. This requires new conceptions of knowledge for human movement studies.

I will return to the issue of reconceptualising human movement knowledge later, but for the moment it is important to recognise that part of the problem is that we are living in a period of economic rationalism in which faculty budgets are under serious review. In this crisis period, education and other social services are under severe criticism, and science, despite powerful critiques of its methods and underpinning ideology, continues to be championed as the sine qua non of progress to a better world.[7] This fact, as von der Lippe explains (chapter 3), has important implications for faculty, curriculum, and the profession in general; and, as such, it is also a political matter.

Professional Missions

Given that much has changed since the times of Dudley S. Sargent, and that we are now in the postmodern era, is it still appropriate to think of the field of human movement as embracing a service mission? If so, whom should human movement professionals serve? What assumptions underpin their mission?

Performance-Oriented Missions

By my analysis, professionals who work in sports science and coaching serve mainly the elite athlete and the institutions which support competitive sports. Although some of them might claim that the knowledge which they develop and implement can also be useful to the ordinary person, their argument is similar to that offered by the motor racing professionals who claim that the sport of motor racing has positive spin-offs for the average motorist. Even if that argument could be sustained, we must return to the issue of the primary purpose of their work—their professional mission.

The clients of the sports scientists, on the whole, are not disadvantaged individuals or groups (Brustad, chapter 6), nor are they "ordinary" people. Rather, the increased commercialisation of sport and the marketing of the body and physical activity since the 1960s has turned sports science into a tool often serving the institutions which benefit from treating the body as a commodity (Fitzclarence 1985; Whitson and MacIntosh 1990). In the commercial world, there is no doubt that performance matters. It means money, research grants, sponsorship, prestige, and national pride. But with such spoils come some related problems; problems which are often ignored or marginalized in professional practice and in the training programs for future professionals.

Certainly, if one wants to improve sports performance, the knowledge which is available within the sports sciences is abundantly useful. However, just as Western medical science does not have all the answers for medical problems, neither does sports science (which is embedded in similar scientific discourses) have all the answers for human movement. In fact, some of the answers of sport science must be reevaluated, especially when referring to issues of personal and social justice.

For instance, although some exercise physiologists may claim to be service- and participation-oriented, their claims are mainly in relation to "health" and its purported relationship with physical activity—a tenuous assumption that fitness *equals* health. Interestingly, as the field of human movement has appropriated the scientific medical discourses, it has warmly embraced the ideology and discourses which advocate this link between exercise and health, while ignoring those messages which reveal that exercise often carries health costs to the individual (Fahlberg and Fahlberg, chapter 5). Thus, in the absence of a critical perspective, exercise physiologists fulfill the role of contemporary exercise evangelists, similar to the physical educators of Dudley A. Sargent's time. The difference is that, instead of the ill health caused by industrial scourges, today's exercise physiologists claim to cure the ills of the hypokinetic and heart diseases of the twentieth century.

Participation-Oriented Missions

Certainly, at face value at least, most teachers and recreation workers for the aged or the disabled work to enhance the life circumstances of ordinary people. In this sense, they are service professionals. Their professional mission, is concerned with the therapeutic and educational dimension of the movement culture. Of course, there are some teachers, and presumably some recreation workers, who, like Andrew Roberts, believe that performance enhancement is central to their mission; but, by and large, even they find, as Andrew did, that such a focus is mismatched to the *needs* of their "clients."

NEEDS AS SOCIALLY CONSTRUCTED. It is sometimes the case that "needs" are created by professionals themselves. As Lawson (1993a) has revealed in his account of the contribution of Dudley A. Sargent, much professional work is predicated on the assumption that professionals (experts) know what is best for people, and that these "experts" alone have the necessary knowledge to help the client satisfy those needs. Human movement professionals have been guilty of this form of paternalism.

According to Lawson, we must "remember that the economic livelihood of a profession—especially its quest for an occupational monopoly—cannot be divorced from its quest for cultural authority" (1993a, p. 11). Cultural authority is achieved when a profession's definitions become most people's definitions. In other words, when it becomes the common definition. He argues that the medical profession has been particularly successful in capturing the cultural authority with respect to health. Health, is essentially defined as the absence of disease, and it is in the interests of the medical profession and its corporate allies such as the drug and insurance industries to so define it. In the case of physical education, we may well scratch our collective heads to isolate what it is we claim and what definitions and problems we regard as our own.

In the Australian context, the cultural authority within the field of physical education is being contested by sports and health "educators." We have seen school physical education defined essentially by its contribution to "health" and most recently by its contribution to the development of sport skills (Tinning et al. 1994). Still, physical education teachers have been unsuccessful in developing the necessary cultural authority to secure an occupational monopoly. When physical education in schools is largely defined in terms of its contribution health or sport, then the claim that only trained physical educators can provide the service in schools and other movement contexts is severely undermined.

The story of the hydrotherapy group in Northcote is an example of clients' needs preceding their participation in physical activity sessions. Interestingly, Marj, the seventy-nine-year-old instructor, would probably not fit into most current definitions of a professional because she did not possess the appropriate degree or professional qualification which certified her knowledge as "legitimate." For me, the success of her classes raises important questions about what knowledge is most appropriate, and for whom.

Towards a Mission for Human Movement in a Postmodern World

So far, I have argued that the context for our professional work has changed significantly since the early days of our profession. In

relation to teacher education, Liston and Zeichner (1991) claimed
that teacher education programs:

> ... can serve to integrate prospective professionals into the logic of
> the present social order *or* they can serve to promote a situation
> where future professionals can deal critically with that reality in or-
> der to improve it. (p. xvii)

Such a claim can equally be made of human movement degree programs. But dealing critically with reality in order to improve it
takes us back to Chomsky's (1989) concerns over the lack of informed public discourse on issues that really matter. A rather brief
scan of the evolution of the field of human movement from its roots
as a serving profession in physical education at the beginning of the
twentieth century to its multidimensional contemporary forms, reveals that there are some enduring discourses which influence
much of the professional practice in the field. Despite compelling
calls for a radical change, or at least for a more encompassing and
appropriate one in accordance with the postmodern times, these enduring discourses continue to underpin performance-oriented practices. They privilege science, place the client in a subservient
position to that of the professional (the professional knows best),
and define knowledge in increasingly fragmented, specialised, and
compartmentalised ways. They characterise modernist Enlightenment thinking.

Scientific discourses can help us understand how to improve the
physical performance of athletes, but are of little help with issues relating to participation in the movement culture. Science, and the
performance discourses should not be the dominant knowledge paradigm for our field. Given that our field is concerned with issues of
participation, as well as performance, it is essential that we accord
proper status and recognition to the social sciences in our professional education programs. But we need more than just more social
science. We need to *reconceptualise* the nature of how we engage the
sciences and the social sciences. Ingham's (chapter 10) conception of
the cross-disciplinary orientation to professional training, Maguire's
(1991) argument for the need of a new multidisciplinary focus on
human development rather than human performance, the human
development perspective advocated by Lawson (1993a) as a replace-
ment to the present human capital model, Fernández-Balboa's critical pedagogical approach to PETE (1993b, chapter 8), and Schwager's
(chapter 9) emphasis on the moral issues of teaching physical edu-

cation are some useful reconceptualisations which we may want to consider. At stake here are crucial questions:

1. Will we continue to advocate the regulation of peoples lives "for their own good?" A democratically encoded agenda advocated by Ingham would have serious implications for our professional power. It would mean a different form of public service from that of human service professions based on a human capital model.

2. Will a broader conception of research be adopted? Recent considerations of the benefits of non-positivistic research for certain questions have been a useful beginning, but still the dominant research questions of the profession remain positivistic in nature. This raises serious questions relating to: (a) what knowledge is worth researching and disseminating? (b) who will control such knowledge? and (c) whose interests will it serve?

3. Will the preparation of educated professionals seriously engage the issue of what exactly is an educated professional? Will such preparation equip the professional to critically engage current practice in such ways as to make it more socially responsible and emancipatory?

One thing seems certain, we must do a better job of initiating future professionals into a more multidisciplinary study of human beings "in the round" (Maguire 1991). Moreover, if we are to lay a legitimate claim to being a human service field, then we must heed Sage's (1993) advice and begin to take seriously the question of whether our professional practices actually contribute to making a better world. We must actively critique social practices relating to physical activity and the body which are oppressive and unjust. We need to explore new knowledge and ways of thinking and be less concerned with guarding our old ideas (Schempp 1993).

Creating or facilitating a socially critical physical education in schools and teacher education is not, however, simply a matter of increasing the number of courses on the social sciences in professional training courses. The evidence is already clear (e.g., Swan 1993; Macdonald 1992) that there is a strong resistance to the discourses of the social sciences by most human movement undergraduates. The point is that, other than for forms of brainwashing, we *do not* have ways of making student teachers think as we might like them to think (Tinning 1993b, 1995). We cannot *make* them think that the issues which we believe to be central to the mission of the profession should be central to them also.

In this postmodern era, a socially critical physical education is certainly long overdue, as is a socially critical human movement profession. Doing something about creating a socially critical physical education will require more than good teaching skills, a strong knowledge of the biophysical functioning of the body, and a love of physical activity and sport by the teacher. It will require a *rethinking* of the nature of school physical education and teacher education (PETE) in the postmodern world. It will require that physical educators be equipped with understandings of the world beyond the gym and the outfield. It will require that teachers be able to see their mission beyond the mere transmission of knowledge about the workings of the body while recognising that their claim to an important contribution to society as a serving profession must rest on the educational rather than the functional aspects of human movement. In sum, as Sage (1993) says, it will require professionals who consider social reconstruction as a significant mission of their work:

> . . . all of us—administrators, sport scientists, and teachers—need to become active agents in contesting dominant social discourses while joining those day-to-day struggles to other actions promoting a more progressive, democratic, socially just society in the wider social area. (p. 160)

Closing Remarks: A Way Forward

One way in which we can begin to rethink our field is to consider the potential of poststructuralist theory and analysis. According to Agger (1991) poststructuralism and postmodernism are overlapping theoretical conceptions. Whereas postmodernism is a theory of society, culture, and history, poststructuralism is a theory of knowledge and language.

A poststructuralist analysis can help us understand the broad concerns articulated in postmodernism. It can help us because it works to make discourse visible. It helps us to "see that our 'eyes have been crafted,' and the political effects of this" (Davies 1994, p. 19). For example, "Poststructuralism has begun to disrupt the binarisms through which we structure our knowledge of ourselves and the social world" (Davies 1994, p. 8). Certainly in the dominant discourses of human movement, the binary is ubiquitous. Human movement is understood in terms of: trained/untrained, football/ dance, fit/unfit, exercise science/history and philosophy, fast/slow, technically correct/technically incorrect, public/private, thin/fat, male/ female, body/mind, and so forth.

Even the heuristic binary presented in this chapter (i.e., performance/participation) would, and should, be subjected to deconstruction by a poststructuralist analysis. Put another way, such an analysis should consider who is speaking (about the knowledge), from what position (of privilege, of epistemology), in what context, and with what political effect. It should provide "insight into the multiple possible readings" (Davies 1994, p. 35) of concepts our profession considers as "defined." To paraphrase Davies for the human movement context, if we are to take the "poststructuralist turn," we must begin with ourselves, our knowledge, and our language, reflecting on how our professional subjectivity has been constructed.

If we are to navigate towards a socially critical human movement that renders its own construction problematic, then it will not be achieved through the application of our previous rational enlightenment (modernist) thinking. In addition to beginning the process of a poststructuralist analysis of our field, we also need to dare to imagine alternatives to the current (limiting) practices of professional training, socialisation, and induction. The following chapters offer valuable practical alternatives in this regard.

CHAPTER 8

Physical Education Teacher Preparation in the Postmodern Era: Toward a Critical Pedagogy

Juan-Miguel Fernández-Balboa

Introduction

When the late tennis player Arthur Ashe was once asked whether his Wimbledon victory was the happiest day of his life, shaking his head, he replied that his happiest one had been the day Mandela was freed. Ashe knew that the highest accomplishment in sport cannot equal human freedom and justice. Yet, it seems that the profession of human movement, as a whole, neither shares Ashe's ideals nor is concerned with matters of freedom and justice. This lack of concern may be due to the influence of modernistic philosophy and the traditional technocratic orientation of our profession. As a profession, we are still operating under the premises of the industrial era and still following the original guidelines which people such as Dudley A. Sargent outlined for us more than a hundred years ago (Lawson 1993a). Times have changed since then, and our society has also changed. We now live in the postmodern era. Then, should we not, as a profession, have changed as well? If achieving freedom and justice is one of the main projects of the postmodern era, should we not, as a profession, aspire to achieve them, too?

Defining Education in the Postmodern Era

"Education" must not be understood as a taken-for-granted, stable concept, but as an ambiguous term whose meaning depends on

121

the ideological lenses through which it is perceived. To help us understand this better, it may be useful to see education as a text (Gore 1990). As with any written text, the concept of "education" can be read, interpreted, and acted upon differently depending on where it is located socially and historically and on the ideological and cultural orientation of the reader. "It is also helpful to locate this notion of text in relation to notions of language and discourse" (Weedon 1987, p. 41). Hence, becoming aware of multiple educational "readings" is essential for understanding how education works and what goals it proclaims to achieve.

Up to now, education has reinforced modernistic values and has been used to promote capitalism and exclusive, elitist notions of culture. It has served to perpetuate the boundaries of power and to separate the dominant from the dominated classes. Yet, given my personal circumstances, history, and ideological position, I believe that education (and physical education) can be a source of social justice and freedom, and its ultimate intent should be to create a better world.

> Education is the point at which we decide whether we love the world enough to assume responsibility for it and by the same token save it from the ruin which, except for renewal, except for the coming of the new and the young, would be inevitable. And education, too, is where we decide whether we love our children enough not to expel them from our world and leave them to their own devices, nor to strike from their hands their chance of undertaking something new, something unforeseen by us, but to prepare them in advance for the task of renewing a common world. (Arendt 1961, p. 196)

Today, fewer and fewer people can deny the urgency to renew our "common world." The modern model will no longer do. We only have to look around us to see how the gap between the rich and the poor has widened; how discrimination in terms of gender, race, sexual orientation, ableness, physical ability, appearance, and religion (just to name a few) has become ever more scathing; how violence and war are still constantly present in our lives; how pandemics and hunger are decimating entire populations; and how the Earth is being systematically destroyed due to greed, ignorance, and irresponsibility. To a great extent, all these are human problems. We, humans not only suffer them, but also have created and contribute to worsen them. In fact, these problems are due to a modernistic ideology, a "preferred reading" (Gore 1990), that circumscribes our ways of "being in the world" to unproblematic technological "progress," to

a never-ending crave for new economic markets and natural resources, and to acts of mindless consumerism. Oddly enough, such
an ideology is being legitimized and reproduced through education
(Giroux 1983).

If we recognize the connectedness between education and society, if we recognize the need to renew our world, then we must also
admit that an alternative pedagogy geared toward creating a society
in which humans live in harmony and respect nature is sorely
needed—a type of pedagogy in tune with the postmodern times.
Through such a pedagogy we can become more civicly and politically
minded and strive for freedom and justice. Critical pedagogy is that
type of pedagogy. Hence, in my opinion, human movement educators have the choice (if not the moral obligation) to practice such a
pedagogy as a source for (a) developing personal and collective consciousness, (b) providing the necessary emancipatory tools, and (c)
engaging in personal and social transformative action (Fernández-
Balboa 1994).

Critical Pedagogy in Physical Education
Teacher Education

Since its conception, physical education teacher education
(PETE) has not been a vehicle for developing students' civic and political attitudes. By "civic attitudes," I refer to a willingness to contribute purposefully and positively to the development of a better
society for all. By "political attitudes," I mean the ability to use
power to retain or obtain control of vital resources and basic human
rights (definition adapted from Bacharach and Lawder 1980).

To develop the civic and political attitudes so vital for having a
free and egalitarian society (Bergen 1994; Callan 1994; Stack 1994;
Ranson 1990; Ranson and Stewart 1989), one must situate the practices in classrooms and gymnasia within larger structures of power
and privilege and, in the interests of great social justice, disrupt the
authority of these conventions by bringing to the fore issues of
power, race, class, gender, physical and metal ability, sexual orientation, and so forth. Alas, one must subject the conventional "commonsense" to scrutiny and critical analysis, examine the forces that
sustain the status quo, and construct new avenues for integration
and possibility.

At stake here are the concepts of intellectual and corporeal lit-
eracies which historically have been supported by scientific rational-

ity and have privileged the white, able-bodied, ecto-mesomorphic, young, middle-class men and, in some cases, women (Gore 1990). Such concepts of literacy can be seen as sources of cultural and physical capital (Bourdieu 1986) because, to a great extent, one's knowledge and physical ability serve as commodities that can be exchanged (as money is) to gain access to, and climb the ladder of, success and power in our society (Shilling 1993b). Those who are not "literate" in such a sense are radically excluded from gaining such access. But illiteracy does not only refer to not knowing how to read, write, or move. From a critical perspective, illiteracy "is also fundamentally related to forms of political and ideological ignorance that function as a refusal to know the limits and political consequences or one's view of the world" (Giroux 1988a, p. 63). Consequently, to become civicly and politically active and to be able to move courageously and consciously toward justice, becoming critically literate is absolutely crucial.

Critical literacy requires that students and teachers learn to interrogate the traditional discourse, personally and socially, from a perspective that is both historical and futuristic. In many cases, it entails rejecting traditional matter-of-fact knowledge. In physical education, this can be done by problematizing our traditions and basic assumptions (Lawson 1993b) and creating more inclusive and equitable alternatives.

Literacy Orientations in Physical Education Teacher Education

Historically, literacy in physical education[1] has centered around the practice of physical activity and the development of sports skills, per se. Such an orientation is viewed as neutral and apolitical. Yet, far from being so, this orientation supports the interests of the powerful groups by reproducing and legitimating unequal relations of power. Fernández-Balboa (1995), Kirk (1993), McKay, Gore, and Kirk (1990), and Tinning (1991) among others have explained that most of what goes on in PETE adheres to the technocratic and performance-oriented ideology. Basically, the orientation of many PETE programs could be boiled down to the concept of "educating the physical," for its own sake—teaching knowledge about sport and physical activity and rendering it beneficial for all. According to Tinning (1991), in this type of pedagogy, "finding the most efficient means to achieve a particular end becomes the dominant issue of concern. The desirability of the end is not contested and ceases to be an important issue" (p. 7). Thus, the problems posed and the questions asked with regards to education are "life adjusting" rather than "life enabling" (Lawson 1993a).

When "reading" physical education in PETE in technocratic terms, several assumptions are made. One of these assumptions is that the teacher (or the professor) is the sole owner of truth whose role is to inculcate "commonly accepted" knowledge to students (Cohen 1988). Another assumption is that knowledge can be taught unproblematically. This is because it is seen as static, neutral, and universal (void of any political or ideological charge). Yet another assumption is that teaching can be reduced to behavior modification strategies, to management routines based on manipulating stimuli that provoke attitudinal changes in students. Put another way, teaching is perceived as a matter of conditioning and manipulating students (Combs 1981).

Within this educational orientation, the main core of study for prospective physical educators consists of technical skills of sport and physical activity, traditional classroom management strategies, and the basic principles for planning and organizing instruction. The process of such training is usually autocratic and individualistic—students of teaching are taught to follow commands and to fulfill the assignments imposed by the instructor. Moreover, most of the instruction takes place out of context—not in the school or the community but in the university's gymnasium, classroom, or teaching laboratory.

From this perspective, teacher expertise is measured by predetermined parameters that center not on the needs of students or society but merely on the technical aspects of teaching. With a few exceptions, teacher educators in these programs neglect social factors and systemic problems:

> For many practitioners of physical education at all institutional levels [including teacher education], social critique is not considered to be particularly important or relevant to what they do or, while they may consider such analysis to be important, is not work they themselves need to carry out . . . is not even on the agenda of considerations. (Kirk 1992b, p. 1)

Under this technocratic orientation, reflection about teaching physical education centers almost exclusively on what teachers and students do in the classroom, seldom considering the implications of such educational practices on both the personal realm and the broader schemes of society (e.g., the cultural, the civic, the political). Reflective practices in technocratic PETE programs are limited to utilizing "pre-programmed" teaching analysis instruments such as PETAI (Phillips and Carlisle 1983), ALT-PE (Siedentop 1991), or some variation of Anderson's (1980) coding system. From this perspective, transmitting conventional knowledge, not social transformation, is what matters most.

Hence, it is little wonder why technocratic PETE programs fail "to produce teachers who have a critical insight into their role and function as teachers in schools, of the value of knowledge they teach, and of the role of schooling in society" (Kirk 1986a, p. 155). Education is considered just a means to securing a job, not a way of challenging students to think more deeply about themselves and the world. Once hired, many graduates of such programs become sullen and silent functionaries with schizoid and passive attitudes and little desire and/or power to change things (Lawson 1988). That is why most physical education teachers cannot be considered professionals in the true sense of the word, just mere technicians.

Beyond the technocratic approach to PETE, there exists a different version aimed at "engaging learners in self-directed and cooperative learning and promoting self-esteem and supportive intergroup relations" (Bell and Schniedewind 1987, p. 55). In this type of PETE program, educators attempt to help learners integrate the emotional and the intellectual aspects into the learning process. The ultimate goal for "humanists" is the students' achievement of self-actualization—a psychological and physical state in which one is able to utilize his/her best talents and capacities (Maslow 1954).

Humanistic educators assume that individuals know what they need to learn (Combs 1989; Rogers 1961). Hence, learning is framed in terms of personal meaning. Meaning, in turn, is discovered through an affective process by which learners experience "reality" and relate events to the self. This way, instead of stressing task-oriented performance, behavior manipulation, and uncritical accumulation of facts, as the technocratic approach does, humanistic programs create a warm and safe learning environment, focus on both the process and the content of learning, and adopt students' own experiences as the foundations for reflection and understanding. Moreover, humanistic teacher educators facilitate students' discovery of personal meaning by encouraging them to take responsibility for their own learning, express their feelings and needs, and resolve personal conflicts. Communication and dialogue are keys in this process.

It could be said that humanistic PETE programs promote education "through" the physical. Yet, despite the value of this educational approach, the learners' education is confined to personal levels; the emphasis on social transformation is usually left out. In these programs, physical education and sport serve as a means to clarifying personal values and achieving self-actualization. Humanistic PETE programs encourage prospective teachers to take responsibility for, and reflect upon, their own learning, to express their feelings, and to grow personally. One can find this type of humanistic experiences most commonly in ropes courses and outdoor adventure programs, although the humanistic philosophy is not limited to these environments. In the city of Chicago, for instance, Don Hellison and his colleagues have developed several successful programs that have a strong humanistic flavor.[2] These programs are addressed mainly to disadvantaged (i.e., mostly poor and non-white) inner-city youth.

A third version of literacy is found in programs whose educational practice is based on critical pedagogy. This approach views didactic tasks and curriculum content as problematic and sees the role of educators in relation to society and the natural environment. To this end, teacher educators and students engage in critical reflection and analysis with the intention to uncover the dominant ideologies; deconstruct taken-for-granted knowledge, meaning, and values; and practice pedagogy as a means for human agency and civic and environmental responsibility.

The goals of PETE's critical pedagogical practice often transcend the classroom and gymnasium walls and enter the realm of the larger community. Civic-minded and political activism are common place in these critical pedagogical sites, and students are encouraged and given numerous opportunities to dialogue, critically reflect, and act in this regard. Those who teach through critical pedagogy in PETE programs, take "teaching the physical" and "teaching through the physical" as a basis for analysis and critique.[3]

Because literacy and curriculum are intimately related (Kirk 1992c), critical PETE programs also examine curricular theories, including those related to the hidden curriculum (Bain 1990b; Fernández-Balboa 1993a; Kirk 1992a). In this vein, as the "hidden curriculum" becomes uncovered, diverse forms of injustice and inequity are not only made explicit but also are contested. Students in these programs learn how "the physical" is constantly used to establish unequal relations of power, and how sport and physical education serve to perpetuate the dominance and privilege of certain elite groups. The intention is not only to raise awareness about issues such as discrimination, racism, sexism, elitism, and so forth, but to struggle against them.

Applying Critical Pedagogy in Physical Education Teacher Education

Lawson (1993b) has urged critical pedagogical theorists to proceed beyond rhetorical opposition to, and negation of, the dominant discourses and provide practical tools to "arm" practitioners with "new knowledge and skills for practice" which will enable them to "fight the battles" for justice and equality (p. 159). In what follows, I will suggest ways of practicing critical pedagogy in PETE through the examination of (a) curricular themes and (b) methodological alternatives.[4]

Examining Curricular Themes in PETE
from a Critical Pedagogical Perspective

To help PETE students reflect about both their role as teachers and the connections between schooling and society, one must attempt to raise their awareness and skepticism about "factual" knowledge. Below, are some curricular themes which students and I often examine and problematize.

REFLECTING ABOUT ETHICS, VALUES, AND MORALS. I believe that clarifying personal and societal ethics, values, and morals is key to developing critical educators (Schwager, chapter 9). According to Bain (1993), "Every choice of what to teach and how to teach involves ethical decisions . . . [that] affect the lives of students in ways that may be right or wrong, good or bad, just or unjust" (p. 70). I believe that critical pedagogues are agents of change, and as such they must be able to act without/against institutional and social coercion. But acting in such a fashion requires the ability to make ethical and moral decisions. Hence, helping students develop moral and ethical skills is crucial (Götz 1989). In order to do so, I create exercises and provide readings that facilitate the reexamination and reflection about values, ethics, and morals. The goal is for students to define what alternatives and actions are best for them so that the values they adopt and the choices they make may be both conscious and intentional, not based on tradition or uncritical inertia.

Becoming a member of the physical education profession means joining a historical "community of practice with a *telos*, a general purpose that one must be committed to in order to be a professional" (Soltis 1986, p. 3, italics mine). Nevertheless, the traditional "virtues" of such community of practice, may not concur with one's ethical principles, and, therefore, a moral conflict with one's sense of professionalism may arise. In fact, considering our archaic professional principles and the new times in which we live, one would expect moral and ethical conflicts. We must understand, however, that conflict is not necessarily bad; in fact, as Fullan (1982) points out, conflict is "fundamental to successful change" (p. 91).

When reflecting about ethics, values, and morals one must think of (*a*) the ways in which knowledge is learned, constructed, produced, sought, and transmitted; (*b*) the multiple purposes, particularities, and subjectivities of teaching depending on the situation, the circumstances, and the context; and (*c*) the possible actions to take depending on one's ideas and lived values (Cutforth and Hellison 1992; Watras 1986).

REFLECTING ABOUT EXERCISE AS HEALTH BEHAVIOR. Because our professional tradition is based on biomedical and scientific modes of analysis, positive health effects (e.g., reduced risk of heart attack, a healthy-looking body, stress reduction) are automatically associated with exercise (Fahlberg 1993; Fahlberg and Fahlberg, chapter 5). In my classes, I provide students with frequent opportunities to question and reflect upon the sources of this information, the applicability of such affirmations, and to what extent these claims are true. By engaging in such an examination, many realize that these "truths" are only partial, and that exercise can be an unhealthy and hazardous practice as well (Koplan, Siscovick, and Goldbaum 1985; Shepard 1984).

PETE students may investigate, for instance, how exercise can be unhealthy when practiced violently, without enough recovery time, or when it is undergone in a polluted environment (e.g., in a exhaust-contaminated city). They may also reflect on how exercise can also be associated with eating disorders such as bulimia and anorexia nervosa (Burckes-Miller and Black 1988; Katz 1986; NEDIC 1988); how it can become an addictive practice (Dishman 1988; Kagan and Squires 1985); and how it may be an oppressive practice (e.g., subscribing to the socially constructed models of physical appearance and attractiveness) (Fahlberg and Fahlberg, chapter 5; Dykewomon 1983—with regards to "fat oppression";

Cherniin 1981—in relation to "obsession with slenderness"; and McNeill 1988).

Following this line of inquiry, it is also important to consider the ethics of health promotion (Fahlberg 1993; Veatch 1982), asking, for example, what are the ethical implications for policy and practice when "prescribing" exercise as a means of coping with stress (Pollock 1988). Moreover, students could investigate the beliefs and attitudes of exercises versus non-exercisers (Slenker, Price, Roberts, and Jurs 1984) and question to what extent prescribing exercise for, say, depressed or obese people is treating the symptom rather than the problem. Also, learners could reconsider exercise and its effects as a sign of broader gender related problems, identity and role adjustment difficulties, and low self-esteem (Nixon 1989).

REFLECTING ABOUT MEDIA REPRESENTATIONS. By learning how to read the media's texts and images critically, PETE pupils can realize how these representations influence individual and social practices (Miller 1990). In order to do so, students can reflect on how the media portray women's and man's images differently. Burton-Nelson (1994), Creedon (1994), and Messner (1988), for instance, have studied this issue as it pertains to TV images of women in sport. Articles by these and other authors could serve to sensitize students to gender discrimination by the media. Racial discrimination issues can also be analyzed by paying attention to how the media portray images of various ethnic groups differently by silencing, omitting, and/or misrepresenting some groups, and highlighting and overemphasizing others. For instance, due to false media representations, many people may be inclined to think that every Asian person is a master in martial arts, or that Blacks cannot swim. This is because members of the former ethnic group are usually portrayed in their "kimonos" whereas members of the latter group are seldom portrayed in the pool.

Students could also analyze the ideological aspects behind media portrayals of "active" men and women (e.g., McNeill 1988), read journal articles such as "Effects of weight training on the emotional wellbeing and body image of females: Predictions of greatest benefit" (Tucker and Maxwell 1992), or investigate alternative images held by women about being healthy, say, from a feminist point of view (Woods, Laffrey, Duffy, Lentz, Mitchell, Taylor, and Cowan 1988).

Moreover, by critically examining TV commercials (Tinning and Fitzclarence 1992) and pregame shows (Wenner 1989), newspaper and magazine photos, and text related to sporting events, students may become sensitized about the media's manipulative strategies.

A good example is Wenner's (1989) analysis of the political "fantasies" and patriotic exultation created in the Super Bowl 20 pregame show broadcast by NBC in 1986. Examining media representations of the body is another example of how to apply critical pedagogy in PETE. "The body," as a construct and as a commodity, is considered very important in our field, and, as such, it has political and economic implications (Griffin 1993; Harvey and Sparks 1991; Sabo and Messner 1993; Vertinsky 1992). We ask why certain body types and colors are attached more prestige and power than others. We also ask in what ways, in our consumer-oriented society, the logic of the marketplace has contributed to creating an image of the body as a commodity—as an object (Fitzclarence 1990; also see Ingham, chapter 10). Plastic surgery, body building, beauty pageants, a perception of beauty linked to youth and thinness, the constant comparison of the human body to a machine and of the human brain to a computer, and so forth, are all examples of this compulsive attitude resulting from the ideology of consumerism. Thinking critically about the body and viewing it from other perspectives (e.g., that of an elderly woman—see Vertinsky and Auman 1988; that of a "handicapped" person—see Patrick and Bignall 1984; that of an Asian person—see Walsh 1983; or that of a society that may be considered "uncivilized"—see Morgan 1991) may help PETE learners realize that there exist other conceptualizations of the body that reflect a more accurate and encompassing picture of reality and that reclaim the body as an integral part of being human (Theberge 1991; Vertinsky 1992). These alternative conceptualizations of the body are more in synch with the postmodern project and, therefore, are worthwhile and necessary for future PETE graduates to learn.

REFLECTING ON DISCRIMINATION ISSUES. Discriminatory images are not only seen through the media. Educational environments are sites where discriminatory practices are common (Fernández-Balboa 1993a). Helping PETE students reflect on some of the practices taken for granted in gymnasia and sport fields is essential for developing better teachers. This could be done by encouraging students to reflect on the effects of having segregated classes for boys and girls (Griffin 1985) and on how members of diverse groups have unequal access to school athletics according to their socioeconomic, ethnic, gender, sexual, physical and mental ability, religious preference, and so forth (Evans 1993; Sage 1990). In our classes, we also think about how the discriminatory practices of sport and physical activity are replicated in other contexts (e.g., the workplace), and we try to come up with effective strategies for contesting and overcoming discrimination.

REFLECTING ON THE "SCIENTIZATION" OF PHYSICAL EDUCATION AND SPORT. Among others, Bain (1988; 1990a), Hellison (1988), Kirk (1986b), and Whitson and MacIntosh (1990), have denounced the dominant influence of the scientific positivistic paradigm in the discourse of inquiry in physical education and sport (see also Brustad, chapter 6). These scholars raise a set of questions that provide us with "ways of thinking that do not take the value of high performance sport for granted and [with] languages that construct other versions of human purposes" (Whitson and MacIntosh 1990, p. 48). Bain (1990a) has argued that in order to construct a world which celebrates ambiguity and competing discourses (i.e., a postmodern world), "we must have new visions and new voices" (p. 9).

In the last few years, we have seen a slow but steady increase in other types of inquiry in our profession, ranging from the interpretative to the critical (Sparkes 1992). Many graduate programs in PETE have adopted and designed qualitative research courses applied to physical activity. Tinning (1992a, 1992b) and others have reintroduced action research as a valid epistemology and practice that can be used to transform our profession into a more socially conscious one. In the same line, critical ethnography (Quantz and O'Connor 1988) has been utilized to examine professional dialogue, and to find the lost voices of the silenced and disenfranchised (Greene 1993). Engaging in these and other forms of inquiry and examining nontraditional kinds of questions can help PETE students and faculty not only become more aware and critical of the issues, myths, and silences implicit in physical education and sport, but also find new solutions for our professional problems.

REFLECTING ABOUT THE NEXUS BETWEEN PHYSICAL ACTIVITY AND THE NATURAL ENVIRONMENT. As a final example of the kinds of critical themes that can be explored in PETE programs, I want to comment on the relations between physical activity and ecology (Zeigler 1986). In fact, most physical activities can be practiced outdoors, and this circumstance may account for the often positive association of the two in the minds of people. Notwithstanding, the practice of certain types of physical activity can have very negative effects on the environment, and vice versa. To illustrate, I will speak of mountain biking and golf. Those with a minimum of sensitivity toward environmental issues can easily imagine the devastating short- and long-term effects (e.g., erosion, life-cycle breaking) that uncontrolled, massive mountain biking may have on local vegetation and the fauna. Often, bikers have little awareness about the dismal impact of such an activity not only on small grasses so vital for the maintenance of the life-cycle and the countenance of the soil, but also on many animal species that have their permanent or temporary habitat in favorite mountain biking areas. Also, many bikers seem to care little about trashing and polluting these areas. PETE students must be educated and prepared to teach others to care for and preserve Nature.

Another example of environmental disturbance and decay can be found in golf. When thinking critically about the ecological costs of this physical activity, one cannot deny the extravagant and unnecessary waste of natural resources golf courses require for their maintenance. Hundreds of thousands of gallons of precious water and tons of dangerous pesticides and herbicides are utilized in each course every year. This not only has direct negative effects on the local habitat (e.g., fauna, flora, humans); but, indirectly, also pollutes and depletes underground water supplies, rivers, and lakes in the areas far from the courses (Tsutomu 1991).

PETE programs can foster critical thinking on issues like these by connecting human relationships with the nonhuman world, seeing Nature not as something to be utilized and abused, but as an integral part of who we are (Brower 1994). By raising awareness about "the special interactions that enhance understanding of the 'self-in-community' with other humans and the biosphere [and the atmosphere]" (Kurth-Schai 1992, p. 150), critical pedagogues can help break down old ideologies about dominating Nature. If we love our ecosystem, if we want to save our planet, we need to go "beyond humanity to embrace all aspects of the biophysical world [thus creating] . . . 'a new way of being human [and practicing physical activity] on this planet' " (Kurth-Schai 1992, pp. 154 and 158).

Examining Methodological Practices in PETE
From a Critical Pedagogical Perspective

In addition to examining content from a critical perspective, critical pedagogy also calls for an alternative methodology based on democratic and egalitarian relations of power. This can enhance students' emancipation and civic-political thoughtfulness. In this light, PETE students and teachers, alike, need to critically revisit the dominant educational precepts and practices through open, honest dialogue and inquiry. Next, I will present some methodological aspects which could be considered. Elsewhere (Fernández-Balboa 1995), I have described these in more detail.

TEACHING PRACTICES. Historically, teachers have been operating from a position of privilege and autocracy (Dippo and Gelb 1991), determining what and how students learn and by when. Many teachers still provide knowledge unproblematically, in the form of lectures and assignments, and take disciplinary action (e.g., lowering students' grades) if students do not comply. Unfortunately, these practices are designed to "domesticate" students and promote their blind adherence to the official subject-based curriculum. As such, students are dehumanized and rendered empty containers to be filled up (Freire 1985).

These oppressive forms of instruction and management are replaced in critical pedagogy by empowering, democratic, and student-centered practices. As such, students are encouraged to take leadership and ownership in the learning process; jointly pose questions and problems; determine (or at least critique and suggest) the course content; and apply knowledge to personal, social, and political contexts. Here, learning becomes humanizing and emancipatory.

COMMUNICATION IN THE CLASSROOM AND GYMNASIUM. In pursuing humanness and emancipation, dialogue is essential (Freire 1985). Contrary to the traditional classroom and gymnasium dynamics in which students are consistently and, often forcefully, silenced, the critical pedagogical forum promotes dialogue and encourages students to engage in "authoring" (Tappan and Brown 1989). This enables them to open themselves and share their goals, needs, opinions, ideas, fears, and so forth. "[By] opening ourselves as imaginative, intuitive, feeling, thinking beings, we may discover something about what it signifies to create our own meanings along the other creatures" (Greene 1978, p. 31). Dialogue enables learners to discover patterns of interaction and come up with insights not attainable by any one individual (Senge 1990). Moreover, respectful and meaningful dialogue is basic for exercising democracy (Fernández-Balboa and Marshall 1994).

By dialoguing amongst themselves and listening to one another, by sharing their stories and helping each other recover their "voices" and true identities, by speaking up and even "talking back" (hooks 1989), PETE learners may gain new and more empowered identities, realize their commonalties and differences, and discover that they all are worthwhile persons. This can be a strong first step toward personal empowerment, which in turn may later lead towards the liberation of self and others.

CRITICAL REFLECTION. Kirk, (1986b) has argued that "educators who lack the capacity for [critical] reflective thought and informed critical judgment may be in danger not only of confirming their low professional status, but also of leaving themselves open to political manipulation . . . [This may also lead to] engaging in simple-minded indoctrination of pupils" (pp. 155–156). To prevent this, and implicit in the idea and practice of critical pedagogy, critical reflection is needed.

Being critically reflective goes beyond having the ability to think seriously, to recollect after thought, to justify what is said and done, and to engage in what Mehan (1992) would call "active sense making" (p. 1). Although these are indeed characteristics of the reflective process, as I see it, they do not constitute critical reflection in a total sense. Critical reflection also requires dealing "consciously and expressly with the situations in which we find ourselves" (Dewey 1934, p. 264). Therefore, the critically reflective act implies and requires an intentional state of "wide-awakeness" (Schutz 1967) as well as an attitude of full attention and thoughtfulness. In addition, reflection does not imply thinking about the past or present only (i.e., about what was or is), critical reflection is also a way to discern "what could be." Ergo, it is "[linked] to the idea of project—the project by which a person, any person invents herself or himself" (Greene 1991, p. 17). As such, reflection becomes a means for transforming the present and for inventing the future. Reflection for transformation cannot be perceived in individual terms exclusively, for the individual is not an isolated being. Instead, it must be seen in relation to one's social and environmental contexts. Only then critical reflection can acquire its full meaning.[5]

GRADING AND TESTING PROCEDURES. Presently, standardized and empirically "objective" testing and grading procedures are still utilized in many PETE programs. These procedures are designed to sort students out on the basis of their performance. Furthermore, testing provides the teacher with a tool for controlling students. Under these circumstances, it is little wonder why these procedures breed anxiety and undermine individual self-esteem and group cooperation among learners.

Kohn (1994) has called for the "[abandonment of] traditional grading and performance assessment practices [in order to] achieve our ultimate educational objectives" (p. 38). This is seen in critical pedagogical assessment procedures, as they are empowering and validating. From this perspective, "accountability" focuses not on what the teacher says students must learn, but on a jointly planned outcome between the teacher and the students. "Students must be invited to participate in determining the criteria by which their work [*sic*] will be judged, and then play a role in weighing their work against those criteria" (Kohn 1994, p. 40). This way teachers and students can share common educational goals and develop a more equitable relationship. Furthermore, by enabling students to take responsibility for their own assessment, they also learn to deal with ethical and moral decisions.

Periodic and regular self-assessment, peer assessment, and individual or group portfolio presentations are examples of alternative grading procedures. In my own graduate classes (Fernández-Balboa 1995), each student presents to the class (orally and in writing) her/his accomplishments, and grades her/himself at the end of each evaluation period. In these instances, although myself and the other members of the group give constructive feedback to the student, the ultimate decision about the grade lies with each individual. In this process, individual freedom and responsibility, on the one hand, and group support, on the other, are crucial. From my experience with this approach, I can say that, more often than not, powerful and compelling learning experiences result from it.

SOCIAL AND POLITICAL TRANSFORMATIVE ACTION. Mainstream education in teacher education is seldom concerned with social and political action; critical pedagogy is. Critical pedagogy goes beyond raising awareness and demasking reality, and enters the realm of political action. As Freire (1985) points out, "the demasking of reality that is not oriented toward clear political actions against the same reality simply lacks sense" (p. 157). By including political action as an inherent characteristic of teacher education, "the teaching act is no longer the endpoint of the process of teacher education but instead . . . [a] point of departure for developing in the student a deeper awareness of the phenomenon of education" (Kirk 1986a, p. 238). Teaching, itself, no matter what orientation it subscribes to, is indeed a political transformative act.

In critical pedagogical settings, students are encouraged to engage in social causes and be agents of change (Sage 1993). Social and political action can take many forms. For instance, PETE students and educators could affect change by promoting more egalitarian and beneficial exercise behaviors, fighting against discrimination in physical education and sport, and raising environmental awareness. This can be done by writing letters and articles denouncing issues of injustice as they relate to physical activity, organizing peaceful demonstrations (for a guide on how to organize this type of demonstrations see, Schmidt and Friedman 1986), and utilizing diverse physical and artistic forms to foster equity and environmental conservation. Here it is essential that teacher educators serve as models of social and political activism.

Closing Remarks

We seem to have entered a new era—a postmodern era. Postmodernism represents a shift from modern ideologies and opens up the possibility for new challenges and hopes. In our PETE classrooms and gymnasia await the future generations of this new era's teachers, and, as Arendt (1961) said, we must provide them with "their chance of undertaking something new" (p. 169). Critical pedagogy can provide these prospective educators with such a chance. In my opinion, the education of the physical per se, although important, is suffused by the precarious social and environmental conditions we face. Paraphrasing Earl Kelly (in Combs 1981), I am convinced that this world can get along a great deal better with a "klutz" than with a bigot, and, like Arthur Ashe, I believe that justice and freedom are far more important than any victory in any sport.

If we accept the postmodern challenges and hopes (e.g., including alternative social, political, and environmental agendas), if we acknowledge the need to renew our common world, and if we agree that human movement professionals can contribute to this process, then we can begin by adopting "critical pedagogy" in PETE programs. Through critical pedagogy, PETE learners can become aware of the profession's limitations and possibilities, learn to demystify the presumed neutrality of physical education and sport, and begin to see the implications of their roles as educators and citizens. Through critical pedagogy, PETE students and teachers can raise questions such as: Who benefits from the actual structure and knowledge of physical education and sport? What oppressive social forms does physical activity perpetuate and legitimize? How can we transform our teaching practices so that our classroom and gymnasia become spaces for social and environmental renewal? Answering these questions can help bring to the surface the political and moral nature of our profession and create new avenues for reform.

There is no doubt that attempting to reform education is a monumental task, and in view of the difficulties of such an endeavor, one could easily give up hope and let one's spirits become pessimistic. My own experience as a critical pedagogue has taught me much about the risks one runs when challenging the *status quo*. Most schools and universities subscribe to technocratic principles, and being critical is often seen as being negative and counterproductive. Notwithstanding, choosing the critical path for me is a positive sign, a sign of caring, a sign of hope. I believe that rendering hope and building communities of difference (Tierney 1993) are fundamental principles of teacher education. I am encouraged and comforted by the idea that teacher education in physical education can contribute to creating a better future. As the Italian activist Antonio Gramsci would say, we must not lose hope, for with the power of our will, we can always make things better. It is in this light that I choose to think of what is possible. This can be our greatest gift to the postmodern generations.

CHAPTER 9

Critical Moral Issues in Teaching
Physical Education

*Susan M. Schwager**

What might our nation reasonably expect of its teachers? First
and foremost, we might reasonably expect that they be men
and women to whom we would comfortably entrust our children. That is, those who choose and are chosen to teach might
minimally be expected to meet the moral criteria we apply in
selecting our baby sitters: that they be models of deportment
and character.

—Goodlad, *Teachers for our Nation's Schools*

Introduction

This chapter is about teaching physical education and about how
we prepare physical education teachers. Its premises are that teaching
is essentially a moral activity and that teachers have moral responsibilities for the nurturing and well-being of those whom they teach.
The following questions will be addressed: What is meant by the term
"moral" in the context of education, and more specifically in the context of physical education? What are the implications of the teacher's
role as a moral steward? How do we prepare teachers who can address
the moral responsibilities inherent in their role? What would morally
responsible physical education look like in a postmodern era?

*I am grateful to Sarah Doolittle, Juan-Miguel Fernández-Balboa, and Hal Lawson
for their comments on earlier drafts of this chapter.

139

Defining Morality

The term "moral" has different meanings. Some define moral in a religious sense as God-given principles of right and wrong (Rachels 1993). Within this interpretation of morality, for example, one tells the truth because God commands it. Morality is also seen as standards of character, as having the capacity to know what is good, and the ability to do good (Rachels 1993). In this sense, telling the truth is desirable because it is what a good person does. In addition, the term "moral" means a way of guiding one's conduct by reason (Rachels 1993). Given this, if a society placed no value on telling the truth, there would be no point in asking questions for, if one had no reason to believe that the answer would be true, attempts to communicate with another would be futile. Telling the truth is, therefore, viewed as the reasonable thing to do.

However, any discussion of what is moral usually includes the issue of whether imposing one's particular set of morals is right or wrong. Hence, if education is morally driven, whose morals should teachers adopt? According to Rachels (1993):

> Morality is at the very least, the effort to guide one's conduct by reason, that is, to do what there are best reasons for doing—while giving equal weight to the interests of each individual who will be affected by one's conduct. (p. 13)

Here, Rachels looks at morality from a perspective of reason, and does remove the concept of morality from the customs of any particular culture or society. However, I think that the view of morality in the context of education goes beyond reason. As members of a particular society (in theory, a democratic one) and of a particular culture we must establish moral imperatives that support the democratic ideals.

Defining Morality in the Context of Education

Interpreting morality in terms of the underlying purposes of education in a democracy brings up issues such as equality and the preservation of justice (DeLorenzo 1994). Teachers are responsible for far more than the dissemination of content. Thus, the moral implications of teachers' decisions are related to the overarching goal of educating children to be productive and contributing members of society, specifically, of a democratic society (Purpell 1989; Goodlad 1994).

In particular, our children need to be prepared to accept the *responsibilities* that accompany the *rights* of democratic living. Understanding how to function within, and contribute to, our political and social democracy will not come from listening to lectures on government and politics, or receiving a passing grade in social studies, but from engaging in the democratic processes (Shor 1992).

Morality as a means for social justice, involves issues related to equality, fairness, and freedom (DeLorenzo 1994). In addition, since democracy may be described as a moral quest, the privileges and responsibilities of citizens must go beyond getting out to vote on election day. Responsible citizens care for each other and participate in their communities in responsible and constructive ways. Responsible citizens recognize the importance of dialogue (Fernández-Balboa and Marshall 1994). Responsible citizens struggle to balance individual rights with the needs of the community. Therefore, the children in a democratic society must be taught the importance of people's interdependence and must learn to accept and support the rights of others.

In view of this, it is reasonable to suggest that part of a teacher's job (regardless of subject matter taught) is to prepare his/her students to be morally responsible citizens. However, dictating to students what is "right" and what is "wrong" as if these principles were content, may not be appropriate. In fact, imposing one's authority and discipline could undermine the principles of democratic community and social responsibility that teachers seek to develop (Giroux and Freire 1989).

Moral Imperatives for Teachers

Goodlad (1992) has identified five domains from which moral imperatives can be postulated. (1) The "personal domain," in which individual meaning comes from experience; (2) The "instructional domain," which refers to the context of the classroom including the teachers' instructions, materials, student groupings, and so forth; (3) The "institutional domain," which alludes to the larger context of the school including scheduling, course offerings, graduation requirements, and so forth; (4) The "societal domain," which deals with expectations for schools, teachers, and students as determined by community and social agencies such as school boards and state education departments; and (5) The "ideological domain," which represents ideas and alternative perspectives (i.e., what *could* be, rather than what *is*).

In addition, the responsibilities of teachers to promote social justice imply two overarching moral imperatives that cut across the five domains suggested by Goodlad. These responsibilities teachers have for providing their students with equity and safety reflect the moral imperative to actively support the rights of all students to learn and to have equal access to knowledge. Moreover, teachers are responsible for providing a physically and emotionally safe environment in which students can learn.

Furthermore, there are issues related to the definitions of equality and equity and the implications these terms have for education that bear examination. Whereas the term "equity" is often used as a synonym of the term "equal opportunity," Evans and Davies (1993) make a clear distinction between the two. Equal opportunity is typically used to describe equal access, and may not in itself be sufficient to bring about equity. Equity, on the other hand, involves making judgments concerning the extent to which a particular situation is just, and acknowledges that merely following a set of rules or laws may not be sufficient to insure justice. For instance, it would not make sense to grant a physically impaired person access to physical education classes without providing him/her with the adequate systemic support. Similarly, providing equitable experiences in physical education requires that teachers examine the program (activities taught, teaching methods used, equipment, facilities, etc.) in relation to the needs of their students.

In teacher education, defining what is meant by a "safe environment" in physical education implies examining issues related to both physical and emotional safety. We (teacher educators) have traditionally done an adequate job of preparing teachers who are conscious about safety, particularly in terms of the physical environment. For the most part, however, this emphasis on "safety first" has been framed in terms of avoiding possible liability for student injury, rather than in a moral sense. The notion of providing an emotionally safe environment has been, I believe, neglected in the preparation of teachers. Occasionally, articles for practitioners will highlight the unsavory impact of some commonly accepted practices (Schwager

1992; Williams 1992), however, there is a need to frame the discussion of safety within the context of the teacher's moral responsibility for the well-being of students. Teachers should be held, and should hold themselves, morally responsible for the well-being of the children in their charge. At the very least, teachers should treat all children equitably. In addition, all children should feel safe in school. The fact is, however, that not all schools provide safe and equitable environments.

Goodlad's five domains provide a useful framework for examining the moral imperatives I have identified within the context of the school. For example, if educators believe that encouraging dialogue among students and teachers is important for helping students understand the importance of individual rights and responsibilities in a democracy, then there is a moral obligation to provide for dialogue among students and teachers in the schools. This moral obligation could be reflected in the five domains suggested by Goodlad (1992) (see table 9.1).

Table 9.1
Domains of Moral Imperative and Dialogical Pedagogy
(Goodlad 1992; Fernández-Balboa and Marshall 1994)

Domains	Expression of the Moral Imperative
Personal	Teachers believe in the importance of promoting dialogue with and among students in school settings.
Instructional	Teachers utilize instructional strategies that promote dialogue in the classroom.
Institutional	Teachers and Administrators contribute to curricular changes that reflect the infusion of dialogue within course offerings.
Societal	School board and state education department members encourage and support the inclusion of dialogical pedagogy in schools.
Ideological	Stakeholders (teachers, administrators, school board members, parents, students etc.) participate in meetings and forums addressing the implementation of dialogical pedagogy in the schools.

Teaching for Social Justice: Postmodernism, Moral Stewardship, and Critical Pedagogy

Postmodernist thinking challenges us to reflect on the ways in which we view our work and ourselves. It challenges us to examine what we know in light of our individual perspectives. Moreover, postmodernism has a strong political orientation encouraging one to reconsider the meanings of power and authority in society. Adopting a postmodern perspective then requires serious introspection regarding the distribution of power and authority, while bringing to the fore critical dialogue concerning how to educate our children to be empowered participants in a democratic society.

A central question related to the purposes of education as suggested by Purpel (1989) is this: Should education serve to perpetuate the culture as is, or transform it? Those who have adopted a critical educational perspective view the purposes of education as transformative and believe in the ability of all people to become educated (Purpel 1989; Giroux 1981; Giroux 1988a, 1988b). In the context of education, the concept of social justice is linked to morality as well. Critical educators must examine daily, from a moral standpoint, what they do and how their actions impact their students (Giroux 1981).

In addition, teachers must be, in Goodlad's words, "stewards of best practice" (Goodlad 1990). That is, besides fulfilling daily teaching responsibilities, teachers need to regard themselves as "change agents" working to transform the programs in which they teach in order to meet the needs of all children.

> It is reasonable also to expect teachers to be responsible stewards of the schools in which they teach. They and they alone are in a position to make sure that programs and structures do not atrophy—that they evolve over time as a result of reflection, dialogue, actions and continuing evaluations of actions. Teachers are to schools as gardeners are to gardens—tenders not only of the plants but of the soil in which they grow. (Goodlad 1990, p. 44)

Moral Physical Education

School physical education programs provide, for the most part, opportunities for students to engage in a variety of activities (sports, games, dance, gymnastics, etc.). These programs commonly cite development of skill, fitness, and good sportsmanship as desirable outcomes (Jewett, Bain, and Ennis 1995). What is taught, and more so, how it is taught, is often determined by the teacher.

If teachers are to promote social justice among their students, then they must simultaneously model socially just behaviors (e.g., treating students fairly and equitably), and teach their students to behave in a socially just manner (e.g., fostering student dialogue concerning what is just, encouraging students to explore justice in the context of their own experiences in physical education and elsewhere).

Promoting social justice in a physical education setting can be done, and is being done, in a variety of ways. For example,[1]

1. At the beginning of the school year, students participate in setting the rules and consequences for acceptable and unacceptable behaviors in their physical education class. This can provide students with an understanding of how certain codes of conduct ensure safe and equitable participation.

2. Students engage in strategies for resolving disputes that arise during game play. This may show them how to monitor their own conduct and thereby help them exercise their own responsibility.

3. Regarding a particular task, students choose ways to successfully complete it (e.g., to choose the level of difficulty with which they feel comfortable). This may provide a valuable new sense of personal accomplishment and self-esteem.

Preparing Morally Responsible Teachers

Just as teachers can be expected to fulfill their responsibilities as moral stewards, teacher educators should be held similarly accountable. Yet, after having spent the past several semesters experimenting with ways of encouraging the preservice teachers in my classes to look at physical education programs and practices in light of promoting social justice in the schools, I can attest to how extremely difficult this is.

In fact, recently, one of the more frustrating semesters I have spent as a teacher educator mercifully came to an end. I had struggled all semester with students in a sophomore-level elementary physical education class called "Movement Experiences in Elementary Schools." Throughout the semester, but especially as I graded their final exams, I realized that despite all my efforts they just didn't "get it." The "it" relates to the moral responsibilities these students would have as elementary physical education specialists in order to ensure their pupil's physical and emotional well-being. I sought advice from colleagues during the course of the semester, and we discussed different strategies I could use; however, nothing seemed to help. A colleague with whom I regularly shared my frustrations assured me that I was doing all I could, and that the problem was with this particular group of students, not with my abilities to teach.

Reflecting on the events of the semester, however, lead me to conclude that the failure was indeed, at least partially, mine. The students didn't "get it" because I hadn't presented "it" in a way that was understandable to them. Part of the problem was my unawareness about how strong my students' experiences and perspectives toward physical education were. Virtually all of the students in this particular class were athletes and they seemed to approach teaching physical education in the same ways they had been coached. This became evident during discussions of what and how to teach children in a physical education class because the students emphasized the importance of teaching "the fundamentals." From their perspective, children needed to be taught the "correct" way of performing particular sport skills as early as possible. What seemed to fall on deaf ears was that teaching skills and movement concepts must be done in a way that respects the individual abilities of each child in their prospective class (Graham, Holt-Hale, and Parker 1993).

There are important lessons to be learned from this experience. The first has to do with the importance for teachers, and in this case, teacher educators to understand the perspective of the learner. The "apprenticeship of observation" is well documented as a strong socializing influence on those who aspire to teach, and I knew this. However, I had not realized the strong influence that my students' experiences as athletes had on their perceptions of their responsibilities as physical education teachers.

The second lesson relates to the need for addressing the moral responsibilities of prospective teachers in a variety of ways. Preservice teachers need to talk about issues related to their moral responsibilities toward their students and toward the schools in which they will teach. They also need to see morally responsible teachers at work. In addition, they need to learn pedagogical skills that allow them to exercise their moral responsibilities in effective ways. I was not able to foster or model these.

My experiences with this particular class highlight some of the practical problems one may encounter when introducing concepts such as the moral responsibilities of teachers, especially when preservice teachers have not previously reflected about teaching from a moral-social-critical perspective. In this vein, Fernández-Balboa (1994) proposes three "cyclical and interrelated" stages: (a) developing awareness and thoughtfulness, (b) providing critical skills, and (c) fostering transformative action.

Developing Awareness and Thoughtfulness

Heightening the awareness and thoughtfulness of preservice teachers with regards to equity and the impact that curricular activities may have on students' feelings can be encouraged by having them examine two constructs: the "hidden curriculum" and the idea of a "normative order." On the one hand, research on the "hidden curriculum" has highlighted the importance of the subtle messages conveyed by teachers' language and behavior (Bain 1976, 1985b, 1990b; Fernández-Balboa 1993a; Kirk 1992b). Studying the "hidden curriculum" and its potential impact on what students learn in school could sensitize preservice teachers about the importance of the nature of communication among students and between students and teachers, especially in terms of unintended and undesirable ideological messages.

On the other hand, Normative Order for Physical Education Teachers (table 9.2) developed by Lawson (1991a) serves as a means for establishing professional standards for physical educators.

These norms have a moral base and can accommodate a variety of curricular models and teaching-learning practices. At the same time, such a normative order can serve to assess different programs. "As in other human service professions, the overarching principles (of the normative order) are moral. One is prohibitive (do not harm clients); the other is prescriptive (place clients' interests above one's own)" (Lawson 1991a, p. 30). Undergraduate curriculum courses often present the preservice teacher with several curricular models for school physical education programs. When using Lawson's normative order model to assess particular physical education programs, it becomes clear that what determines the degree to which any program is morally responsible is not its goals (e.g., develop sport skills, fitness, or knowledge of human movement concepts), but how the curriculum is taught, and specifically, how each student perceives and experiences it.

Table 9.2
A Normative Order for Physical Educators

Moral	Prescriptive	Prohibitive
	Place each individual's needs, interests, and aspirations before selfish self-interest	Do not harm inviduals
	Enhance each individual's development, health, and well-being	Do not discriminate against individuals of different gender, race, ethnicity, religion, sexual preference, physical fitness level, body type, or skill level
	Demonstrate caring and concern for each individual	Do not ignore political, economic, and social consequences of work practices
	Strive for virtue and justice via personal values, thoughts, and behaviors	
Aesthetic	Prescriptive	Prohibitive
	Enhance each individual's enjoyment of, and involvement in, diverse forms of exercise and support	Do not compromise the intrinsic values of exercise and sport
	Facilitate each individual's discovery of personal meaning through various forms of exercise and sport	
Procedural	Prescriptive	Prohibitive
	Involve individuals at appropriate stages in goal setting, content-activity selection, and evaluation	Avoid diagnostic stereotyping
	Personalize instruction and performance feedback	

(continued)

Table 9.2
(continued)

Procedural	Prescriptive	Prohibitive
	Incorporate relevant research-based knowledge and new technologies in practice	
	Engage in continuous reasoned deliberations about the moral implications of work practices and personal lifestyles with regards to fitness and sport	

Note: In this table, I have replaced the term "client" (used by Lawson) with "individual" to reflect a more egalitarian and less hierarchial relationship between teachers and those whom they teach.

Source: Adapted from Lawson (1991a). "Three perspectives on induction and a normative order for physical education," *Quest* 43(1), 20–36.

Providing Critical Skills

The skills that will enable preservice teachers to address issues related to social justice in their teaching can be described in three categories: managerial, instructional, and reflective. Class management and behavior management techniques are routinely taught in the context of "methods" courses, and the importance of these skills should continue to be emphasized. A novice teacher will not be able to promote social justice in the context of physical activities in classes that are unruly and/or chaotic. Siedentop (1991) emphasizes the necessity for effective class management as a condition for learning, and effective physical education teachers know the importance of establishing rules and routines in their classes. Notwithstanding, these managerial practices must be oriented and practiced within moral and equitable parameters.

Beyond moral managerial practices, the two instructional models offered below are relevant to a discussion of the moral dimensions of teaching because they both encourage decision making and self-responsibility on the part of students. Mosston and Ashworth's (1994) "spectrum of teaching styles" has evolved over the past twenty-five years, but the basic premise of examining the dynamics among teacher, student, and task still prevails. I render the "spectrum" a very useful tool for teacher educators. Presenting preservice teachers with even just a sampling of the different styles in the "spectrum" can help open their eyes to the decision-making dynamics of any interaction between teachers and learners. The challenge is to encourage prospective teachers to apply the spectrum of teaching styles as a means to increase students' involvement in making decisions related to their own learning and to see the moral implications of applying the spectrum (e.g., increasing freedom, practicing self-regulation, engaging in democratic decision-making).

By altering the decision-making dynamics in a particular lesson, students have opportunities to make choices about what and how they will learn, take responsibility for their own learning, and realize the implications of doing so for themselves and the group. The spectrum provides a decision making continuum in which the decisions made by the teacher and students can be examined as to whether they meet the needs of the students while being consistent with democratic principles.

Hellison's model designed to develop students' responsibility for their own behavior may also be useful in the process of moral development (Hellison 1985, 1995). The model, which describes students' social behavior in four representative levels is a valuable resource for addressing issues related to the development of social responsibility in (originally but not exclusively) disenfranchised youth (Hellison 1985). The stages in this model (i.e., awareness, problem-solving, student sharing, and self-reflection), do help students become aware of the impact of their behavior on others.

Also, providing undergraduates with techniques for ongoing reflection of their teaching particularly in the ways suggested by Hellison and Templin (1991), can help develop their competence and caring as teachers. The two questions that frame this latter approach ("What's worth doing?" and "Is what I'm doing working?") can be introduced to preservice teachers for them to examine managerial and instructional behaviors in a moral sense. Here, with regards to the moral function of the teacher, one more question should be asked: "Is what I'm doing the right thing to do for the students?"

Fostering Transformative Action

Important first steps for teacher educators are to reflect on how we are preparing our students and to provide opportunities for them to (a) become responsible, and (b) acquire skills that will help them fulfill their responsibilities. We also need to look at what is happening and what can happen in our own classes and programs, while working with other teacher educators, teachers, and administrators in the schools. Action is, after all, the ultimate aspiration. Without transformative action, all I have said above is of little use.

Changing Roles of Schools and Universities

Recent developments in educational thought have the potential of changing the roles of schools and universities in dramatic ways. Here I will offer three themes as starting points for thinking about how we may view schools and the role of physical education programs in the schools, as well as how to prepare physical education teachers. The first two, "nurturing pedagogy" and "simultaneous renewal," call for a reexamination of the mission of schools, and the process for changing them. The third, "service integration and interprofessional collaboration," addresses the fundamental needs of children and their families and calls for a change in the way in which we view our professional responsibilities. This third theme is centered in the notion that our roles as professionals and citizens cannot be separated.

Nurturing Pedagogy

The notion of education as child-centered rather than subject-centered is crucial to approaching teaching in ways that are morally responsible. Traditionally, elementary school programs have been more child-centered than secondary ones. This is evident from the way in which the school day is organized. Elementary school children spend the school day (for the most part) in one classroom with one teacher. Secondary school days are organized by subject matter. Secondary school teachers are more likely to describe themselves as teachers of subject matter rather than teachers of children. Ask an elementary teacher what he/she does, and the response will be "I teach third grade" or "I teach third graders." Ask the same question of a secondary school teacher, and the response is more likely to be "I teach math, or history, or english." Based on my experiences supervising student teachers in elementary and secondary schools, I have found that teachers of the "specials" (art, music, and physical education) at the elementary level also tend to be more child-centered in their teaching than their secondary counterparts.

Noddings (1992) has suggested a different way of viewing the school curriculum which represents a dramatic shift (especially at the secondary level). She proposes organizing the curriculum around centers or themes of care. Specifically, these themes of care would be: "care for self, care for intimate others, care for strangers and distant others, care for nonhuman animals, care for plants and the living environment, care for objects and instruments, and care for ideas" (Noddings 1992, p. 70). As such, acquiring knowledge about science, math, history, english, and so forth, would be accomplished in the context of these themes of care. With regard to physical education, Noddings notes:

> First, it seems that many of the isolated topics we now teach might reasonably be integrated into scientific and technological themes within the care context. I will argue that the courses we now call physical education, home economics, driver education, sex education, drug education, health and hygiene, and parenting ought to be integrated. Teachers who now work in these areas should be encouraged to form one large department that can provide continuous discussion on topics of essential care for the physical self. (Noddings 1992, p. 75)

In addition, according to Noddings, physical education programs, as we now know them would cease to exist:

> Physical *"education"* departments that limit themselves to the supervision of sports and exercise should be eliminated. In the years after high school, no one forces us to exercise or participate in games. We all have to learn why and how to exercise our bodies and to take responsibility for our fitness. Proper physical education would provide open discussion on issues of fitness, monitor the condition of bodies, and prescribe possible modes of exercise and recreation. Secondary school students would then be encouraged to use the various exercise facilities available at schools. Team sports should be offered also, but as part of a complete educational program in which competition and cooperation are discussed and analyzed. (p. 75)

Simultaneous Renewal

Goodlad (1994) has suggested that for schools to change, the renewal process needs to happen simultaneously in the schools and in the institutions in which teachers are being prepared. Simultaneous renewal is possible when educators from the schools and from the universities view themselves as members of the same community. Goodlad offers some specific ways in which schools and universities can work together for their mutual benefit and for the improvement of education. He suggests developing Centers of Pedagogy in which teacher education faculty, faculty from university arts and sciences departments, and teachers and administrators from the schools work together in unique ways.

Goodlad's call for renewal is clearly grounded in the mission of schools as teaching the young the principles of a political and social democracy. He stresses the need for ongoing educational renewal toward achieving this mission and insists that schools and universities must work together for the improvement of education (Goodlad 1991).

Interprofessional Collaboration and Service Integration

There is plenty of evidence that hungry, sick, abused, homeless, disenfranchised children and youth have difficulty learning even in a caring, nurturing school environment. Based on the premises that our society is in crisis and that past efforts to address the mounting problems have not been effective (Stallings 1995), some suggest that schools should be the centers for other social and community-oriented activities (Dryfoos 1994). Recent interest in the integration of education and human services is another child-centered approach that touches on the moral implications of education beyond the "what" and "how" we should teach our children. This is a new model for practice called "interprofessional collaboration," which makes possible a related strategy called "service integration" (Hooper-Bryar and Lawson 1994; Lawson and Hooper-Bryar 1994).

> Service integration involves the creation of "seamless systems" of education, health and social services. The idea is to tailor offerings to diverse kinds of people in their often-unique contexts and integrate them. "Wrap-around strategies" for service integration thus involve blending heretofore separate and competing policies, professionals, programs and supports into a coherent, cohesive, culturally-responsive, and context-sensitive framework. It takes as its point of departure the lived experiences and felt needs of people, and it requires professionals to mesh and integrate programs, supports, and services as these people suggest. (Lawson 1994, p. 9)

The benefits of providing youth and families with the kinds of services they need in order to survive and flourish are morally compelling. The task of providing these services in an integrated "seamless" fashion and in a way that empowers families is daunting. It is clear, however, that colleges and universities need to be involved in these community based, child- and family-centered activities. We must

Susan M. Schwager

begin to prepare physical educators who see education as child- and
family-centered, and see it as their moral obligation to engage themselves as participants in these efforts to provide integrated services
to children. Accepting our responsibilities as professionals to promote social justice and providing moral stewardship means that we
act to ensure the basic needs of children and their families and, as
such, endeavor to create a better society.

Closing Remarks

The postmodern theme of this book calls for dramatic changes in
how we think about our work and our lives, as well as how we act.
Adopting a postmodern perspective gives us the opportunity to critically examine taken-for-granted aspects of our lives and our work.
The so-called postmodern thought provides an important "political
[moral] and theoretical service in assisting those deemed as 'Other'
to reclaim their own histories and voices" (Giroux 1991, p. 24). It
challenges the legitimacy of unjust social practices and calls for a
better society. The theme of this chapter has been the moral responsibility of teachers for the care of children. I encourage teacher educators to prepare teachers who recognize and are responsive to the
needs of children and students. In addition, I argue that the teachers we prepare should accept, as part of their professional role, the
social responsibility of nurturing the democratic ideals and behaviors. Maxine Greene's (1995) description of the teacher's role in a
postmodern world challenges us to put our students at the center of
our actions, emphasizing our moral responsibilities toward justice
and freedom:

> As teachers, we cannot predict the common world that may be in the
> making; nor can we finally justify one kind of community more than
> another. We can bring warmth into places where young persons come
> together, however; we can bring in the dialogues and laughter that
> threaten monologues and rigidity. And surely we can affirm and re-
> affirm the principles that center around belief in justice and freedom
> and respect for human rights, since without these, we cannot even
> call for the decency of welcoming and inclusion for everyone, no mat-
> ter how at risk. Our classrooms ought to be nurturing and thought-
> ful and just all at once; they ought to pulsate with multiple
> conceptions of what it is to be human and alive. They ought to re-
> sound with the voices of articulate young people in dialogues always
> incomplete because there is always more to be discovered and more

to be said. We must want our students to achieve friendship as each one stirs to wide-awakeness, to imaginative action, and to renewed consciousness of possibility. (p. 43)

Preparing morally responsible teachers and promoting social justice in schools are challenging and necessary goals. Physical education teacher educators must take seriously the moral and civic responsibility inherent in our roles as teachers by asking ourselves how we can ultimately meet the basic needs of the children and enable them to grow and participate fully in a democratic society. Change is difficult, but thinking differently about what we do is a beginning. We have the capacity to act as moral agents to change what is not working in the schools and in the communities. Acting as such is what will make a difference.

CHAPTER 10

Toward a Department of Physical
Cultural Studies and an End
to Tribal Warfare

Alan G. Ingham (and Friends)[1]

Introduction

Postmodernism is a time of challenge, of reconstitution, of re-
conceptualization, of reconstruction. Therefore, we must first examine our present constitutions, conceptualizations, and constructions.
In moving from deconstruction to reconstruction, I shall first expose
some contradictions in the current constitution of our field. Then I
will introduce some new domain assumptions, a mission statement,
and a new field focus. Finally, I shall outline some building blocks
from which to construct the new curriculum in departments of physical cultural studies based on a curriculum that is critical, integrative, and in tune with the postmodern times.

The Academic Tribes

I suggest that, historically, three tribes have emerged in our field
(Adams 1988): (a) the technicist practitioner, (b) the technocratic
intelligentsia, and (c) the humanistic intellectual. I also suggest
that, in this historical conjuncture, each possesses what appear to be
nonnegotiable values or exclusive, self-defining value referents. Such
positions provoke tribal warfares and perpetuate traditional sub-
disciplinary agendas and the ghettoization of knowledges. At the
core of all this is the acquisition and maintenance of academic and
social prestige.

157

The Exoteric-Esoteric Dimension of Prestige:
A Weberian and Sartrean Perspective[2]

Gramsci (1971, p. 9) stated, "All men are intellectuals, one could therefore say: but not all men [*sic*] have in society the function of intellectuals." How might we understand this remark in Weberian terms? I believe that Gramsci's idea of people being intellectuals resides in the notion that all strata of society have a common intellectual orientation, that is, *pragmatic rationalism*. Arguably, then, regardless of whether one is a laborer or a professional, one takes a pragmatic orientation to the attainment of immediate goals.

In terms of prestige, in our rationalized world, we work somewhere on the exoteric-esoteric continuum. To follow Gramsci (1971, p. 10), the difference between exoteric and esoteric knowledge is that the former is *technique-as-work* and the latter is *technique-as-science*. The interesting point here is that both exoteric and esoteric knowledges are technical knowledges. On the one hand, labor, especially atomized and deskilled-degraded labor (Braverman 1974), requires little esoteric thought. It is routine and boring. Professionals, on the other hand, are probably respected because they have an esoteric discourse derived from specialized knowledges to which the public does not have cultural access. Language and knowledge combined lead to power and, hence, prestige. Those lacking in such find themselves in positions of dependency.

In the modern conjuncture, the pragmatically rationalist orientation to action is overlaid by *instrumental rationality*. In instrumentally rational action:

> ... the end, the means, and the secondary results are all rationally taken into account and weighed. This involves rational consideration of alternative means to the end, of the relations of the end to other prospective results of employment of any given means, and finally of the relative importance of different possible ends. (Weber 1947, p. 117)

In prestige terms, the ability to engage in the rational consideration of alternatives distinguishes the professional from the laborer. The latter's work rarely requires the exercise of substantial rationality because substantial rationality is disciplined by the formal rationality of production. Mannheim (1940, p. 53) described substantial rationality as:

> ... an act of thought which reveals intelligent insight into the inter-relations of events in a given situation. Thus, the intelligent act

of thought itself will be described as "substantially rational," whereas everything else which either is false or not an act of thought at all . . . will be called "substantially irrational."

Thus, it is not only one's situation in the exoteric-esoteric continuum which distinguishes the laborer from the professional; it is also the occupation's inducements to engage in substantial rationality or, to put it another way, the degree of relative autonomy which is inscribed into the work process.[3]

Where Do We Stand in the Society Wide,
Occupational Prestige Hierarchy?

Traditionally, all professions have one thing in common—public service. In the case of the professoriat, it is the service function of the intellectual function with which the public is familiar. The lay public may, at times, question the value of a tertiary education, especially when the knowledge imparted is esoteric and appears to be noninstrumental concerning vocational training, but people generally understand the teaching function of the intellectual function. Thus, it is the professional-client interaction or the public service function which lies at the heart of the public's evaluation of our status. In most cases, the public turns to professionals for the expert knowledges the latter possess. What the public typically does not understand, even if they possess a bachelor's degree, are the ways in which these knowledges are produced.

Really useful knowledge, for many, is exoteric knowledge which can be applied. Thus, the farther away from application that knowledge production seems to be, the less the majority of the population understands the intellectual function (The Ivory Tower critique). Within the established academic disciplines, however, the further the knowledge production process is from direct application, the more the producers of such knowledge are accorded prestige. At the heart of the matter, then, is an exoteric-esoteric problematic which creates a status ambiguity concerning the evaluation of the professoriat's performance of the intellectual function. The exoteric-esoteric dimension also has implications for the profession / discipline dualism that we, in the human movement profession, have created within our intellectual function.

THE TECHNICIST PRACTITIONERS. The technicist practitioners can be easily recognized. Many of us have been such in the past when we started our tertiary education wanting to be a teacher or a coach. As we progressed through the sheepskin steeplechase, and although still committed to our original pedagogical and didactical calling, instead of teaching in the public schools, we decided to teach teachers how to teach. The prestige of performing our pedagogical mission seemed greater in tertiary rather than primary or secondary education; however, our mission orientation remained the same (i.e., custodial, normative, prescriptional, and reproductive) merely applying and implementing typically politicized policy dictated by bodies such as school boards and state legislatures. Here, student aspirations and legislated knowledges articulate in the function of the technicist practitioner. Accreditation standards or guidelines thus determine what is considered as really useful knowledge when it comes to curriculum offerings and credentialing.

Technicist practitioners are to be found in the professional schools, but rarely in the established disciplines. In our field, the technicist practitioners seldom contribute to the knowledge production process. Typically, they are the victims of intellectual snobbery and are left out of the knowledge production loop. When involved, they become the objects of deductive research performed by the technocratic intelligentsia from within an established paradigm.

THE TECHNOCRATIC INTELLIGENTSIA. I call this tribe technocratic because, despite their objectivistic fantasy, there remains a value-free, pragmatic orientation to knowledge production. Its members tend to teach the specialized, "sub-disciplinary" classes. Moreover, because they are enamored with objectivistic pretensions and hard data, their work typically "statistical neopositivist" and subscribes to the ideology of "mastery" and "control." Their research is basically instrumental, functional, and pragmatic. We usually see the label "applied scientist" affixed to this tribe.

The technocratic intelligentsia's function is located above the technicist practitioner and, hence, one interactional level away from the public seeking professional services. If they come into direct contact with a public, it is usually a public of their own choosing. Here, the public not only receives services, but also is likely to be the source of data as subjects in research. The technocratic intelligentsia's research produces the knowledges needed by the technicist practitioners to educate and certify their students (presuming, of course, that the technicist practitioners read this published research).

For the most part, the technocratic intelligentsia are engaged in what Kuhn (1962) calls "normal science" exercises suggested by the positivistic paradigm. Here, the members of the technocratic intelligentsia appear to have "no choice" but to work from within this paradigm. Given that their specialized knowledges are not relational and contextual, the technocratic intelligentsia initially try to resolve problems at the methodological and operationalization levels. This is a drawn-out process involving replication. However, if one looks at their reviews of the literature, one sees emphasis upon recency rather than primacy. Thus, with time, the original idea loses its resonance and becomes "taken for granted." Critical reflection about the functions, implications, and objectives of the task is seldom considered. Only if, during the mopping up and extending the range process, persistent anomalies are discovered will the paradigm move from pre-reflexive into critically reflexive consciousness. Thus, anomalies usually are surprises which cause crisis in the cultural formation or school of thought.

In terms of numbers, this tribe is the dominant one in our field and in other professional schools. In recent years, members of this tribe have been influential in changing our departmental titles. While our technicist practitioners would have probably been content with a departmental label such as "Health, Physical Education, and Recreation," our technocratic intelligentsia found such a label lacking in communicating their sub-disciplinary intentions and devoid of prestige vis-à-vis their disciplinary colleagues on campus. Those who suffered from natural scientific pretensions, for all the above reasons, found the term "Kinesiology" appealing. Once embraced either in name or in spirit, the writing was on the wall for those in our field who were engaged in the humanities and the soft social sciences. Thus, over the last twenty years, we have seen either the marginalization or the elimination of those who would qualify as humanistic intellectuals.[4]

HUMANISTIC INTELLECTUALS. The humanistic intellectuals are humanistic not because they embrace humanism as a philosophical school of thought but because of their philosophical, humanities, social, and scientific education. The key to understanding the humanistic intellectuals is to be found in the "culture of critical discourse" (CCD). Yet, although it appears to be oxymoronic, there is a kind of reactionary utopianism in their careers. The CCD is based on an ideological culture aimed at promoting discourse among the humanistic intellectuals as well as advancing their cultural capital. CCD is characterized by:

> . . . speech that is *relatively* more *situation-free*, or more context or field "independent.". . . [This] special speech variant also stresses the importance of particular modes of *justification*, using especially

explicit and articulate rules. . . . The culture of critical speech requires that the validity of claims be justified without reference to
the speaker's *societal position or authority* . . . Good speech . . . has
theoreticity. . . . CCD is also relatively more *reflexive*, self-monitoring, capable of meta-communication, that is, of talk about talk; it is
able to make its own speech problematic, and to edit it with respect
to its lexical and grammatical features, as well as making problematic the validity of its assertions. CCD thus requires considerable
"expressive discipline" not to speak of "instinctual renunciation."
(Gouldner 1979, pp. 28–31)

CCD irritates the members of the technocratic intelligentsia.
Humanistic intellectuals demand and search for a context beyond
narrowly defined prescriptive and proscriptive knowledges. Their
willingness to entertain multi-determinations, their insistence on interpretation rather than revelation, and their emphasis on recursive
regularities rather than laws, are at odds with hypothetico-deduc-
tive research and deny preferential status to the natural scientific
orientation and to neopositivistic epistemologies. In this regard, the
humanistic intellectuals are, for the most part, meta-paradigmatic;
although, granted, it is fair to say that they, after many years of critical and self-reflexive thought, have decided to be loyal to a given
school of thought and, oftentimes, will be its advocates. I have tried
to capture the differences between the technocratic intelligentsia
and the humanistic intellectual in ideal typical fashion in table 10.1.

From Burning Bridges To Building Bridges

Having portrayed the tribes in our field in ideal typical terms,
the following question arises: Is there anything we can do to bind
the tribes together around a collective focus and mission—a focus
and a mission which could eventually lead to the elimination of
tribal warfare? It is my intent to engage in a "value-referenced" selection and to posit a curriculum for our future. To this end, I shall
go beyond deconstruction to the construction of a curriculum that is
free of the structural faults that have plagued us professionally and
intellectually.[5]

Towards a Focus and a Mission

"Practices" need to be distinguished from "activities." In our
field, we are accustomed to using the latter. This is because there is
a lot of a-theoretical behaviorism underlying our endeavors. "Activ-

Table 10.1
The Technocratic Intelligentsia
Versus the Humanistic Intellectual

The Technocratic Intelligentsia	The Humanistic Intellectual
time-bounded	context-bounded
novelly reproductive	emancipatory
incorporated	resistant
instrumentally positivist and empiricist	culturally critical and empirical
scientific-technical	philosophical-literary
topical specialist	generalist
textual	contextual
pragmatic	idealist
objectivistic	value-referenced
functionalist	usually dialectical
measurement-minded	concept-minded
politically dependent	politically oriented
tied to economic markets	tied to cultural markets

Note: I have not included the technicist practitioners in this table because, in recent years, they have tended to be left out of curriculum defining and building processes. One might assert that the technicist practitioners are an "actively residual" (Williams 1977) tribe whose missions and values are overlooked by both the technocratic intelligentsia and humanistic intellectuals in recently reformed departments.

ities" can be left in its a-theoretical, behaviorist connotation, but, for my purposes, it is a subset of "practices." Practices are recursive, regularized, and, therefore, "rule-governed" components of human agency (Bourdieu 1979; Giddens 1976). Also, practices are enculturated and social-relational—there is relational meaning behind the doing; and the doing, in turn, makes for relational meaning. Practices are socially recognizable. Think, for example, of the practices which relate to hitting a ball with some sort of a stick. Even if no implements are visible, we can discern and label the differences between someone practicing a golf swing and another practicing a baseball swing or a tennis swing. In other words, we can not only impute meaning because of our shared cultural meanings and formations, but we can practice our whatever swing on the basis of recursive knowledge—we think of the immediate past to improve our present or we recapture our past as a way of evaluating our present. "Cultural meanings and formations" are made on the basis of shared and distinguishable practices. This idea applies equally to the "golfist" and the "positivist." It is through their shared and distinguishable *practices* that our academic tribes can be recognized. Our doings and our reasons for our doings define us culturally.

I think that we have favored "activities" in our vocabulary because we view each other as "human doings" (evaluating each other's relatively discrete activities) rather than "human beings" (seeing one another through empathic emotional resonance based on interpersonal or social-relational practices). The former is why many in our field have selected "performance" as the focus for their units. Those who have led the kinesiological cultural formation seem to be particularly prone to assume that their enculturated individualistic, control-based, and achievement-competence orientation should be reproduced in, and legitimized through, our curriculum. Thus, these individuals regard anything that does not have a performance-enhancement outcome as superfluous.

To rectify this state of affairs we must articulate our mission with our focus. Missions involve goals or general purpose statements which are often ideals in the sense that they may not be immediately attainable. Yet, they provide committed direction. We, as a field, have had a long-standing commitment to the preparation of teachers and coaches. I am not suggesting that we abandon this commitment in our endeavor to seek prestige from those external to our field. Thus, our aim should be to enrich our teacher education programs by adding substantive knowledge about exercise, nutrition, health, sport, and leisure to the knowledge in didactics, and conversely, to enrich our allied professions' programs with knowledge of didactics. After all, many of our allied professionals are involved in instructional practices. However, I insist that we must provide a context to and for the text because human movement practices are referenced to cultural formations (e.g., social classes and status groupings) and to institutionalized arrangements (e.g., religion, politics, mass media) which are not physical cultural or systematically educational. Our mission statement should acknowledge this while, in the form of an organizational charter, identify selective excellence—what we do best.

Hence, I propose creating Departments of Physical Cultural Studies (DPCS). Below, I shall employ what Max Weber (1949) called the ideal-type. An ideal-type is both instrumental and insufficient (Sadri 1992). It has heuristic value and it must be insufficient if we are to avoid the dangers of "reification" (Lukàcs 1923). In other words, an ideal-type is a scaffold; erected to affect, but not to dictate. Therefore, although in utopic terms, I shall use the ideal-type as a measuring rod against which we can evaluate our current practices and those we might produce as we transform the curriculum. Therefore, I offer the following, value-referenced, focus statement for the DPCS:

> The DPCS consists of cross- and inter-disciplinary studies of practices in physical culture (e.g., movement activities, exercise, nutrition, training, enculturation, recreational and representational sport) which critically assess and promote programs that focus upon the intersections of physical activities, health behaviors, and movement-related lifestyle choices.

From this focus statement, a departmental mission statement is in order. The DPCS will:

1. Provide public service through consultation, policy formation and implementation, continuing education, and internships, all of which involve democratically encoded agendas derived from mutual interaction among participants;

2. Based on this grounded knowledge and other knowledges gained through our own cross-disciplined inquiries, develop through research and disseminate through courses, presentations, and publications, scholarly knowledge about practices in physical culture;

3. With knowledges anchored in the above, prepare educated professionals for work in a variety of public and private organizations which have, as a whole or a part of their own mission, the provision of physical cultural services for those in need.

The DPCS will prepare educated professionals and scholars through a curriculum in which those involved will:

1. Acquire a shared language (discourse), knowledge, and skills in order to facilitate collaborative learning, scholarship, and service;

2. Learn how to learn by evaluating different kinds of knowledges; by engaging in critical and self-reflexive practices, and by learning how to produce new knowledge;

3. Gain an understanding of the interrelationships among practices in exercise, health, nutrition, sport, and physical leisure to promote individual and social betterment. Also, they gain insights into the principles and processes which can enable them to understand regularized and recursive practices in exercise, health, nutrition, sport, and physical leisure. (To this end, faculty should be involved interactively in the educational process—for example, curriculum development, teaching);

4. Learn to recognize and appreciate sources of human variability including, but not restricted to, contributions to personal biography made by biology, heredity, race, ethnicity, gender, social class/status, diverse cultures, and political and environmental forces;

5. Learn to recognize and appreciate sources of human commonalty including, but not limited to, physical cultures, the human body, entropies, and disease;

6. Learn to recognize and appreciate how knowledges can contextually and appropriately be understood and applied as well as how knowledges can be misinterpreted, distorted, or applied erroneously;

7. Learn to recognize and appreciate ethical and moral issues regarding programs and policies in health and physical cultures through liberal education and cross-disciplinary principles.

The Orientation of the Curriculum

In determining the DPCS's curriculum, it is important to distinguish between a cross-disciplined orientation and a sub-disciplined orientation. The latter predicates on the notion that one can acquire prestige from those who already possess it. In this sense, prestige is a means of social control and, as a result, requires punctilious conformity on the part of those who aspire to be among the ranks of those who already possess it (Goode 1978). Thus, as we organize ourselves around a focus, either through training in the parent discipline or by imitation of the parent discipline, we introduce into our field the epistemologies, methodologies, theories, and methods of our respective parent disciplines. As such, our respective claims to fame appear to be anchored in being more like those than in being ourselves. In short, the very idea of being sub-disciplinarians is linked to a powerful dependency relationship with *the* academically recognized disciplines and their hegemonic epistemologies.

The cross-disciplinary orientation reorganizes the power relations between the sub-disciplines and the parent disciplines. It does so first, by having a unique and self-sustaining focus of inquiry; second, by moving from a situation of dependency to one of interdependency both *between* academic units (an inter-disciplinary orientation) and *within* our academic unit (the cross-disciplinary orientation); third, and this is why it is a new configuration, by breaking down the epistemological, methodological, and theoretical tribal boundaries. This new configuration of social relations and interdependencies lessens the need to reproduce, internally, the external disciplinary contents and forms that have traditionally compartmentalized the production of knowledges in our field—what C. Wright Mills (1959) called "abstracted empiricism." Moreover, when fully actualized in the curriculum, this new configuration engages faculty in cooperative endeavors which, by virtue of epistemological cross-fertilization, eliminate the knowledge hierarchies typically found in our field. (I shall return to this later.)

From Building Blocks to Erecting the Curricular Scaffold

Above, I laid out our functional mission statement. The curriculum is one important means of attaining it. But how shall we decide on its contents? Throughout this book, we address some of the perennial, structural fault lines in our profession and in the way it communicates its knowledge. In this part of the chapter, I would like to readdress some of them, beginning with our two paradigmatic cultures.

The natural sciences primarily produce knowledge deductively because they are overdetermined by paradigms; whereas, the social sciences, because they are under-determined by paradigms, produce knowledge both inductively and deductively (Brustad, chapter 6). To be sure, this is a gross generalization but nonetheless a useful one. It distinguishes inquiries into those that are "presuppositioned" and those that are "presuppositionless" (Willis 1980). This distinction is artificial,[6] but, in our field, it constitutes a serious cleavage and becomes the source of invidious comparison between those who do "science" and those who do "studies." Of course, the distinction is also specious, especially as it applies to our field. The paste-pot eclecticism that is characteristic of research in our field is the result of an instrumentally positivistic orientation which embraces both the natural and social scientists in a belief in incrementalism—that one day, and as a result of our nominalist endeavors, the big theoretical picture will emerge.[7] Here, even our symbolic interactionist and participant observationist colleagues have an immanent compact with both functionalism (Gouldner 1970) and positivism (Willis 1980), because they preserve the subject as an object and retain the notion that the object can present itself directly to the observer (Willis 1980).

There is also another related and serious epistemological cleavage which needs to be addressed. It is anchored in the hermeneutic (von der Lippe, chapter 3). Giddens (1982) argues that the natural sciences have a single hermeneutical problematic; whereas, the social sciences have a double hermeneutical problematic. "Hermeneutics"—the science of interpretation—has long been ignored in our field mainly because our quest for prestige caused us to draw upon the positivistic philosophies of the natural sciences.[8] This means that even though the natural science method might legitimately treat the object of study as an object that does not answer back, yet, when it comes to policy, our objects of analysis are, in truth, subjects of analysis who do answer back and who are rule-governed not rule-bound. Policy is, after all, *socially constituted*. In this regard, one does not explain the practices of human beings (i.e., discover laws); one has to interpret these practices through the interpretations of those engaging in them (i.e., discern regularities and recursiveness and their underlying meanings depending on the context and the circumstances).[9]

The recognition of the double hermeneutic and its application in research might yield significant understandings concerning why people do not always do what we think is best for them. Resistance to our prescriptions and proscriptions does not simply arise out of ignorance (the education ideology) or laziness (the anti-welfare state ideology), it also arises out of local, popular cultures and economic status. I should also add that we should stop teaching in a non-hermeneutically informed manner. In treating our students as objects, rarely do we involve them in the knowledge production and acquisition processes nor do we engage them as subjects-as-objects in our experiments in the classroom and our curriculum designs. Yet, what they know might surprise us and force us to question the inflexibility of our texts.

There is a third problem for our technocratic intelligentsia in using presuppositioned practices: The preoccupation with the refinement of statistical techniques and research instrumentation; the endorsement of a nominalist or individualistic conception of society; the affinity with verificationism and incrementalism; the linkage of a (false) dichotomy of facts and values with a (mis)conception of value-freedom; and the prominence of team research and the multiplication of centers or institutes of applied research (Bryant 1985). Put another way, in moving from theory through methods through operationalization to data collection, the scope of the research becomes increasingly narrow in focus and decreasingly holistic. Thus, the policy to which the research applies suffers from the very blinked vision.[10]

It is not my intention to beat up on the natural scientific method and its instrumental positivistic epistemology. They have their legitimate place. My point is that we do not need knowledge hierarchies derived from the "hegemony" of instrumental positivism because it legitimates the work of the technocratic intelligentsia and the technicist practitioners (Ingham 1986; McKay, Gore, and Kirk 1990; and Brustad, chapter 6). However, our mission should dictate our epistemologies and methods, not the other way round. For the most part, our technocratic intelligentsia and our technicist practitioners are unaware that they adhere to this epistemology because they have not been exposed to knowledge derived from philosophy and social-philosophy of science. Rarely do we require such exposure in the graduate education of our students. Indeed, since many faculty are instrumental positivists, they encourage intolerance of philosophy, history, and other forms deriving from the culture of critical discourse (Bryant 1985).

If our mission is, in the final instance, social and physical cultural, then it is logical to propose that our natural science and would-make-social-science-into-natural-science colleagues take the double hermeneutic seriously in their applied research endeavors. Not to do so has resulted and will result in civic a-morality and the absence of professional ethics (Durkheim 1992).

No Ghettos in Our Edifice

Sub-disciplinary knowledges are ghettoized knowledges—"I do this, she/he does that"—even among humanistic intellectuals who uncritically embrace multiculturalist and empowerment agendas. There was a period in North American physical and health education when the hegemonic curriculum was a sport education (fitness being a spin-off) curriculum. The sport education curriculum was (and to some extent still is) a hegemonically heterosexual masculine

170   *Alan G. Ingham*

and Anglo-Saxon curriculum. We expected categories of people who were not Anglo-Saxon, heterosexual men to accommodate their cultures to the dominant one. While there have been oppositional cultural formations throughout the history of our professional field, I think it is fair to say that they gained momentum during and after the civil rights activities of the 1960s. And I think it is fair to say that many cultural formations that pursued the civil rights agenda did so within the moral agendas of liberal democracy.

By so doing, the marginalized extracted concessions from the dominant, primarily on the platform of distributive justice in recruitment and reward (equality of opportunity rather than equality of condition). However, structural assimilation did not eliminate the hierarchical status among the cultural formations themselves. Cultural difference remained a key problematic especially when ranking differences in terms of superiority and inferiority. Presently, the masculine, Anglo-Saxon hegemonic intellectual focus still remains quite strong. However, the groups whose cultural formations fall outside of WASP andrarchy continue to reject its domination and convincingly argue for status equality on the cultural terrain. Hence, the adoption of a postmodernist concept: multiculturalism.

In order to obtain status equality on the cultural terrain, cultural formations that were not WASP andrarchical had and have to wage wars of strategic position. To attack the dominant, counter-hegemonic status groups have parceled out the dominant with respect to their special interests. WASP andrarchical has become atomized into hegemonic white, hegemonic Anglo-Saxon, hegemonic Protestant (although this has declined in salience), hegemonic heterosexual, and hegemonic male. Each of these hegemonics has a counter-hegemonic "-ism." So, instead of engaging in a truly relational form of analysis, the counter-hegemonic "-isms" appeal for distributive justice in categorical terms rather than in relational terms. Thus, we have courses on women in sport rather than gender relations in sport; on African-Americans in sport rather than race relations in sport, and we have created centers which lie outside of departments in order to recognize the validity of counter-hegemonic knowledges. I am not saying that such curriculum developments were not needed. What I mean is that the conjectural moment has passed for them to be needed in their original forms. As originally formed, there were simply too many categorical essentials involved.

To say that such categorical strategies are not needed now is not to dismiss them. It is to advocate the need to find new articulations. For example, the key to a gender relational analysis lies in understanding that gender relations are sexual politics (Carrigan et al. 1985). Thus, demystifying hegemonic masculinity requires us to deconstruct heterosexual politics. Here it is not only women (of any sexual persuasion) who are victimized: it is also effeminate men, gay men and women, the old, and children who are victimized. Victimization is the core of categorical articulation. So, in what we now know as sport, there are plenty of victims because sport has encoded contents, forms, and relations ruled by hegemonic heterosexual adult males. Paradoxically, women who are good at men's sports are viewed to be masculinized, and men who are good at men's sports are never thought to be gay. In both cases, heterosexuality is reaffirmed by dismissing the concrete contradictions. In terms of the sexual politics of complementarity, "real" women should not be in masculine sports, and "real" men should not do ballet. If women are in masculine sports, they are not seen as "real" women. If men are in dance they are not seen as "real" men. By the same token, children in sport are not considered children, but miniature adults playing by adult andrarchical rules which "separate the men from the boys."

Teaching From the Scaffold

From all that has preceded, we can say that instead of encoding the curriculum and continuing to reproduce knowledge hierarchies and sub-disciplinary segregation, we should organize ourselves around the study of practices in physical culture, with their historical, cultural, structural, personal (mental-manual) components interwoven into it. Practices reproduce and transform. Practices are constituted and constitutive. They structure structures and are structured by structures (Giddens 1982). We are departments of physical cultural studies and, in such departments, exercise practices, health practices, nutritional practices, sport practices, leisure practices, and pedagogical practices should not be hierarchically arranged, categorically boxed, discrete academic territories. In such departments, no epistemology should be privileged.

In the same vein, this means that we must reduce our traditional emphasis on instrumental positivism, and view it as a paradigm so full of anomalies that a revolution is needed. There is simply no reason to believe that, because we have had and still have a service function, instrumental positivism is the best way of producing practical knowledge or a knowledge of practices. By deconstructing this emphasis, we will undermine the tribal claims of the technocratic intelligentsia vis-à-vis the technicist practitioner and the humanistic intellectual (Gouldner 1979).

Moreover, by ridding ourselves of our objectivistic claims to be value free, we can admit what we have known all along; namely, that science is value-referenced and value-relevanced (Weber 1949). This admission is especially important to our field because what we do has political, social, economic, and psychological consequences concerning our bodies and our selves. What we can do is to engage in critical and self-reflexive thinking, not as a negative form of thinking, but to borrow from Sartre (1956), as our ability to adopt an external point of view on our selves, on our selves in relation to others, and on our knowledges in relation to other knowledges.

What does all of this dense stuff mean for the DPCS curriculum? First, in our teaching, we have to take Kuhn (1962) seriously. First, as a means of critical and self-reflective thought, we must remove dualisms. Second, in working with contradiction instead of normative functionalism, we can develop a different sense of determination. Instead of simplistic notions of cause and effect, we can become attuned to the notion of multiple determinations and to the idea that determinations are not one-directional. We can do so by removing another of the cornerstone epistemologies of our field. (i.e., functionalism). "Functionalism" is the belief that if something is persistently present, it must have a function (Giddens 1977).[11] If we rid ourselves of functionalism, we can cease talking about levels and start talking about interpenetrations. Third, we must begin to pinpoint contradictions that not only have not been resolved homeostatically, but also have precluded rational-purposive action in the quest to replace existing systems with new ones. Contradictions are more than just conflicts of interest. They typically involve historical disproportions between forces and relations of production. As such, they can become the bases of new interpretations and help us develop a new sense of determination. In our field we have ignored this knowledge and continue to place more emphasis upon functional "needs" and "oughts" than upon contradictions.

Fourth, we must realize that human practices are constituted and, at the same time, are constitutive. They do not merely reproduce, they transform. Thus, any given curriculum, as a cultural practice, might be looked upon as an artifact that is continuously becoming obsolete. But, within the curriculum, there are contents or problematics that are organically durable yet require conjecturally changeable interpretations. The organically durable problematics are the ones we can legislatively tell our students to know. The conjecturally changeable interpretations merely modify our ways of knowing about or "seeing" in the sense that the durable are histori-cized (contrasted in time) and compared (contrasted in historical time and cultural milieus).

Fifth, all of the above requires that we develop a teaching strat-egy which will encourage critical and self-reflexive thinking in our students (Fernández-Balboa, chapter 8). And the only way I can think of doing this is by being critical and self-reflexive ourselves. Here, I would suggest that we present to our students not only the conjecturally "hegemonic" (i.e., power assertions in a given period of our history), but also the active residuals (i.e., what still remains through inertia), the counter-hegemonic (i.e., what struggles against dominance), and the alternative-hegemonic (i.e., what seeks new power structures and relations and envisions "what could be"). This also applies to our presentation of paradigms.

Sixth, because I believe that our students have the right to know where we are located in the knowledge production process, it is time that we admit that we are value-referenced not in the nor-mative functionalist presentation of our mission as what "ought to be," but in our presentation of ourselves as scholars. This exercise in honesty is a valid way to face our "hidden" categorical and self-serv-ing agendas. We may not like what we see and, if we are not willing to make amends by removing our phobias, we may, at least, feel some guilt about exhibiting them in public (i.e., constructing a cur-riculum that reproduces them in our students). A socially responsi-ble curriculum is one that leads ourselves and our students to possibilize the "other" in the "self" and vice versa (Fahlberg and Fahlberg, chapter 5).

Seventh, it matters little if one is dealing with epistemologies, theories, methods, data (contents), or policies. Our students need to know why they are learning what we try to teach them and what al-ternative knowledges and methods are available. They need to be situated in the knowledge production process. They need to know "whose rules count" and why, in order to know something about "power over vis-à-vis power to" and, therefore, to acknowledge their responsibility for their professional practices. I have tipped my hand throughout this essay: Empowerment is not just a matter of rights; it is also a matter of responsibilities.

Eighth, knowing what we are, in generic terms, does not en-able us to declare or possibilize what we can become. To use some fancy structuralist language, we must articulate the articulations. We are not dependent sub-disciplinarians—subunits of the parent departments (e.g., we are neither exercise physiologists nor physiologists of exercise). We are a faculty who have an understanding of how exercise is enabled or constrained by our genetically endowed anatomies and physiologies, our social structural milieus, and our enculturative *habitus*.[12]

Finally, we must acknowledge that we are not departments of leisure studies. Rather, we are a faculty who are interested in the physical cultural practices of human movement and in how these articulate with other practices that are not purely physical cultural per se (e.g., social class, gender, ethnicity, employment/underemployment/unemployment, etc.). In the same vein, we are not Centers for Women's Studies or African-American Studies, and although our knowledge about physical cultural practices may be informed by their work, we are not their logical subsets or their adjuncts. It is therefore important to distinguish our curriculum, in the abstract, from our individually, value-relevanced, scholarly interests. Sometimes our scholarly interests and the curriculum will articulate nicely; other times they will not. We can inform the curriculum with our interests, but hiddenly forcing our personal interests or political agendas to dominate the curriculum would be a mistake. At any rate, if we do share our interests, it must be in the open, so that they can be criticized and scrutinized.

Similarly, given our mission statement, we are not in the School of Fine Arts. Most of what we purvey falls into the realm of the popular rather than either the realm of haute couture or the elitist and nationalistic heritage definition of Culture. Thus, we are not in the performing arts as ballet educationists or modern dance educationists. We are movement educationists, and movement is a generic, skilled, and cultural practice. In this regard we are not sport educationists either. The performing art of sport has its own department— the department of intercollegiate athletics. Moreover, we should not be sport educationists if we are in the business of eliminating categorical differences. For example, women cannot really win—categorically, distributively, or relationally—in sport because the contents, forms, and relations of sport as we now know it are encoded and embedded in masculinity. But does this mean we should abandon sport studies? Of course not! Sport is a significant component of popular physical culture. For many, it is the hegemonic form of physical culture. However, in the role of relational analysts, not of skill educationists, we must also include the recreational, critical, political, economic, and representational spheres of sport. Here we are neither

sport psychologists nor psychologists of sport. We are neither sport
sociologists nor sociologists of sport. We are a faculty which wonders
about the hegemonic sport practices in relation to the broader range
of movement practices and the psychological and social consequences of this overemphasis relative to status group prestige and
resource allocation. In short, we need to link skill in movement to its
biographical aspects, its social structural and cultural determinations, and its "representational/interpellative" consequences and
alternatives.

**Generic Bricks, Conjectural Representations,
and Cross-Disciplined Inquiry:
Contents of the DPCS Curriculum**

Then, how should we decide which problematics to include and
exclude in the DPCS curriculum? I submit that the key to a liberally
educative and cross-disciplined core curriculum lies in finding the
generic problematics, thinking about them in terms of the lived experiences of the students, and interrogating these through the development of cross-disciplinary questions. In view of that, first, we
need problematics which resonate with our students' lived experiences. We are, after all, seeking to contour students' educational experiences through an inductive or grounded-theory orientation to
knowledge construction. Second, we need problematics which are
amenable to cross-disciplinary inspection. Inductively, we must integrate and move towards context in the form of cross-disciplined interrogation. The objective would be to move students and faculty
from pre-reflexive thinking to critical and self-reflexive thinking,
constantly engaging in exercises of problem setting and problem
solving. This way, we could elicit what, how, and why questions from
the students. Our role as collaborative teachers is to ask ourselves:
What would be my contribution to the problem-forming and knowl-
edge-constructing classroom environment? How does my expert
knowledge contribute to analysis and interpretation? How does my
contribution interpenetrate the contributions of students and other
faculty members? Upon what shall I focus my critical and self-
reflexive capabilities and those of my students?

    In this fashion, generic problematics would be continually
posed. However, our attempts to address these problematics should
be conjectural—articulating them not only with the residual (past

knowledges) and the dominant "solutions," but also with the values embedded in the general cultural order. At the heart of physical culture stand the human body and movement.

Critical and Self-Reflexive Evaluations of the Representational Body

I use the word "body" as a totality, not as one part of the dualism mindbody (Fahlberg and Fahlberg, chapter 5). The body is, at the same time, both physical and cultural. It is both an object and a subject. It is genetically structured and endowed with potentials. What we are given, we take up or develop. How we develop and why we develop this or that aspect of the body, how and why we use our bodies to this or that end depend on economic, social, political, and cultural multi-determinations. The object-body is there! We use it; we exercise it; we nourish it; we tinker with it; we expose it; we conceal it; we adorn it, and so on. To say all this is merely to describe how we do and what we do, and the effects that are produced by how and what we do, but there is little interpretation of why we do it. Interpretation becomes necessary when we begin to view the body in its social-political-cultural milieus (Turner 1984). Here, it is essential to address how natural development and social-political-cultural development articulate, and how the person(ality) is formed through these articulations.

In "physical culture," all of us share genetically endowed bodies, but to talk about physical culture requires that we try to understand how the genetically endowed is socially constituted or socially constructed, as well as socially constituting and constructing. In this regard, we need to know how social structures and cultures impact our social presentation of our "em-bodied" selves and how our embodied selves reproduce and transform structures and cultures; how our attitudes towards our bodies relate to our self- and social identities (Fahlberg and Fahlberg, chapter 5); how our body work (diet, exercise, skill acquisition, abuse, and other regimens) connects to the general cultural order; how our socially negotiated body (appearance, manner) or our symbolic or "sign-ing" body signifies itself, and why the signs are significant (Kirk, chapter 4). In short, we need to know how our genetically endowed bodies are variously enculturated and how the various enculturations are interpreted by those who have the power in structure and culture to recruit and to reward.

When we socially and culturally manipulate our bodies (or let our bodies be manipulated), we enter into a relationship with our bodies—their appearances and mannerisms—and into a relationship with others who are having a relationship with their bodies. So as not to appear too voluntaristic, I need also to say that when our bodies are socially and culturally manipulated, we find ourselves in a relationship with others whose bodies have been socially and culturally manipulated, too. Through acts of self-reflexion and social comparison, we can identify ourselves with or distinguish ourselves from these others so producing distinctive yet interpenetrating physical cultures. All of these are examples of practices—regularized and recursive—in physical culture because all practices are relational.

Critical and Self-Reflexive Evaluations of Movement

In addition to the problematics core, we also need to think about the development of a practicum in movement—a practicum which invites the exploration of meaning in and through movement. What does the "doing" of movement signify or represent concerning the *social being* and *identity* of the mover in movement, and how the mover was enculturated?

Movement is a neural-physiological and kinesiological activity, but it is much more than this. It is an enculturated practice which is anchored in "culturated" distinctions that can reproduce or resist the hegemonic cultural order. Our movements then are not simply functional; they are culturally and ideologically encoded and become presentations of the self in the social-relational situation (Goffman 1959, 1981; Fahlberg and Fahlberg, chapter 5). In short, our movements are not always politically innocent. Movements, mannerisms, and appearances communicate things about an individual's social status and "warnings" about the interactional role the individual will expect to play in the social situation (Goffman 1959). We move according to status symbols and wear status and esteem symbols to signify social rank or merit in the rank (Goffman 1951). Such symbols range from wearing a medal to a having suntan in winter.

Mannerisms, for instance, are enculturated and signify not only *status* (e.g., age, gender, social class), but also *intent* with regard to power (e.g., inclusion and exclusion, recruitment and reward). In our hegemonic, masculine, youth-oriented sport, what does it *mean* to be told that "you throw like a girl?" What does it *mean* to be "doddering," and does it *mean* the same for the young and the old? Doddering may meet with different types of relational response depending upon the age of the dodderer. Movements and mannerisms are also involved in the cultural politics of being "cool."

I am sure that you can think of many more examples and conceptualizations of how to critically examine "movement." My point is to get us to think of how to sensitize our students to movement as an anatomical, neural-physiological, *and* political-cultural problematic. We need also to sensitize our students to the ways our bodies and movement forms are rationalized in Weberian and Freudian terms.[13]

In addition, we also need to think about the development of a practicum in human movement—a practicum which invites the exploration of meaning in and through movement. Examples of questions posed in this practicum are: What does the *doing* of movement signify or represent concerning the *social being* and *identity* of the mover in movement? How are movers enculturated? Needless to say, in this context, the movement practicum should be more than a skill learning exercise or a performance evaluation exercise. We already offer such experiences to our students, but we need to reconfigure such experiences into a decidedly physical cultural curriculum. My suggestion is to configure the practicum in much the same way as the core curriculum—sensitizing displays followed by team teaching and cross-disciplinary interrogation concerning contents, forms, and relations.

Theorizing the Problematics: Higher Order Articulations in the DPCS Curriculum

New trends in the development of capitalism are unhinging our once taken-for-granted political economic articulations. They are also challenging traditional occupational categories. Such events have been heralded as the advent of postmodernity (Bauman 1992; Harvey 1989). My intention has been to argue that occupational categories and the sub-disciplinary degree tracks that serve them are now impermanent. This is not to say that the knowledge which we impart is necessarily outmoded, but it is to say that knowledge which is organized along sub-disciplinary tracks (e.g., exercise science, health studies, sport psychology, sport and cultural studies) now becomes a limiting condition concerning the placements of our students under conditions of flexible capital accumulation. With this in mind, existing forms of curriculum tracking will have to be replaced in ways which will allow our students to respond to changing social, political, and cultural conditions without us having to sacrifice the integrity of our knowledge base.

Upper-Divisional Classroom Experiences

My suggestion here is that we move to a modular curriculum beginning in the students' junior year. At this point, we need to think through how much of the curriculum should be legislated and how much knowledge should be interpreted (Bauman 1987). Certainly, we need modules which will provide the foundational knowledges. Rather than tracking, I suggest we allow students to follow their interests across modular offerings in the *basics*. Then we can create the more specialized knowledge clusters also in modular form. These knowledge clusters may be either more structured (legislated) for those seeking a professional credential, or less structured for those seeking a non-professionalist liberal education. However, it should still remain possible for students to make connections with specialized knowledges—that is, articulate knowledges relevant to their preprofessional training for a career. For example, a student with a desire to enter the fitness industry might require knowledges traditionally associated with dietetics, exercise science, health education, leisure studies, marketing, management, and pedagogy. In attempting to make these connections, the student will require more innovative advising that is typically required in a sub-disciplinary, tracking-style specialization.

Furthermore, there is the question of pedagogy-didactics and the professional ethics involved in such. Traditionally, in our departments, we have reserved pedagogical-didactical experiences for those pursuing teacher education. Given the postmodern needs outlined above, reserving such knowledges for the credential-minded seems to be an erroneous strategy. A large proportion of our graduates will be engaged in a wide range of educative-enculturative endeavors. They will be interacting with "clients" and will need to be both knowledge experts and teachers. Thus, we need to create modules which not only are geared to teaching in the classroom environment, but also are geared to teaching outside of the educational system. This way, while the would-be teacher would have a more structured set of knowledge clusters, those in the human movement profession who will be teaching outside of the educational institution would be able to incorporate modules on pedagogy and didactics for the allied professions. I suggest that in the newly created DPCS, such a strategy would not only make more sense for students and future professionals, but also would breath new life into the quite exhausted and disenchanted human movement faculties.

Upper-Divisional, Non-Classroom Experiences

Most of us have experiences with certification programs and internships, so I shall not focus upon these. Rather, I want to introduce another capstone experience which genuinely tests our commitment to cross-disciplinary learning and to our mission and focus statements. It is the Public Service, Team Research Project.[14]

One of the traditional criticisms of university education suggests that it is out of touch with the realities of the world in which we are preparing students to be employed. One way to ameliorate this condition is by making the instructional environment more congruent with the demands and responsibilities of employment. Few of our students want to be like us—professorial. Yet, many will find themselves in various form of professional practice and face professional demands. The Student Team Project (STP) is designed to pre-professionally prepare our students for such exigencies while, at the same time, to demonstrate our commitment to community service.

I have acknowledged that some specialization will occur within a cross-disciplinary curriculum. Hence the STP would begin in the first semester of the senior year, following the students' involvement in DPCS research methods modules. The DPCS faculty, jointly with the students, and following the suggestions of external clients or sponsors, would identify projects relevant to existing problems. The faculty and students would initiate contacts with clients and sponsors and "democratically" encode problems and projects for the student teams to address. Each team would have a faculty advisor, fairly familiarized with the project at hand. The team would act as a consultant for clients or sponsors such as schools, community organizations, health providers, civic governments, athletic clubs, and so forth. What the STP would do is to familiarize students with the process of (*a*) problem identification, (*b*) the establishment of project boundaries, (*c*) goal and objectives setting, (*d*) proposal drafting, (*e*) study or project design, (*f*) data collection, (*g*) data analysis, (*h*) implementation, and so forth.

Closing Remarks

All works seem to require summaries. This one does not. The problem with summaries is that they suggest a point of closure. This work has not such point. Rather, it is a beginning, an intention seeking actualization. It is a "useful fiction" (see Weber 1949) seeking non-fictional status through your various interpretations and applications.

PART II

Critiques of the Critical Postmodern
Analyses of the
Human Movement Profession

CHAPTER 11

Transformation in the Postmodern Era:
A New Game Plan

Linda L. Bain

Introduction

Twenty-five years ago, as head coach of the women's basketball team at the University of Illinois-Chicago, I scouted the game of an upcoming opponent. The team I came to watch was losing, as the other team executed one fast break after another. Sitting a few rows behind the team bench, I could overhear the frustrated coach talking to her team during a time out. As the players sat exhausted from running up and down the court, the coach was angrily exhorting them to become more intense, to try harder. After this scene was repeated about three times, one of the players exploded, "We're trying as hard as we can. What we need is a new game plan!" The player was right.

The social problems confronting our society are frustrating and exhausting. Many of our political leaders are exhorting us to try harder, to do a better job of implementing the principles of the modern era. They propose that what is needed is better science and technology, greater individual responsibility, stronger sanctions for undesirable behavior, and more incentives that reward personal and corporate achievement. The authors of this book argue that such reforms are not enough to create a just society and that radical transformation of the system is needed. The new game plan they recommend is *transformative postmodern theory*.

In this chapter, I will provide a commentary and critique on the alternatives being proposed in this book. I use the plural, alternatives,

183

because the authors do not speak with one voice. An edited book is a social construction in the most literal sense. A loosely connected group of people are gathered by the editor because of their interest and expertise in a particular topic. Each author, with the guidance of the editor, creates a chapter that describes his or her interpretation of the topic. The product is a multivocal construction that reflects the group's shared understandings and also highlights the unique perspective of each author. The authors of this book share certain characteristics; all are university professors of kinesiology (or a related field) with training and experience in pedagogy or social science. All have a commitment to a critical theory stance, but the extent to which the authors are grounded in a postmodern perspective is less clear. Although the authors share some characteristics and assumptions, their chapters reflect differences of nationality, gender, and professional experience.

Each reader of the book also brings experiences, knowledge, and values to the reading of the text, creating another layer of multivocality to its interpretation. Those of us who have been asked to write chapters for the last section of the book have been invited to make our interpretations public. In doing so, we may provide new insights into the content of the earlier text, but more certainly we will reveal something about our selves, our perceptions, and our values. As you read these commentaries, you are encouraged to reflect on your own reactions and to recognize that your reactions are also a social construction and a valuable source of personal insight.

I have struggled with how to organize this chapter. Consistent with postmodern theory, my reactions to the book are layered and multifaceted. Rather than trying to synthesize them into one coherent whole, I have decided to present each layer separately. I begin with a theoretical analysis of the logic of the book's basic arguments because, like most of us shaped by the modern era, my "instinct" is to view reason and logic as a foundation for action. The influence of postmodernism has caused me to question this assumption, but despite my mid-life corrections, my persistent tendency to privilege rationality is reflected by my decision to put logical analysis first. In later sections I will examine the implications for professional practice and personal relevance of the postmodern perspective.

A Theoretical Analysis

The premise of the book is that the overwhelming problems confronting society require not merely more effort but a new strategy, one that would change the way we define the problems as well as our approach to solving them. Juan-Miguel Fernández-Balboa, the editor of this book, lays out the following basic arguments in chapter 1:

1. The profession of physical education is grounded in modernity;

2. Modernity has failed;

3. Western society is moving into a postmodern period that challenges basic principles of modernity;

4. To address the failures of modernity and to survive and thrive in the postmodern period, the profession of human movement needs to change;

5. The redefinition of the profession in the postmodern period needs to have a critical, emancipatory emphasis.

The authors of the subsequent chapters speak to each of these points in differing degrees. I will not attempt to systematically review each of the chapters, but will ask if the underlying arguments for rethinking the profession in the postmodern era are clear and convincing. Yet, to evaluate the arguments, the reader needs to understand the definitions of modernism and postmodernism. The first three paragraphs of the introduction provide a description of the tenants of modernity summarized as follows: "At its center, the image of a coherent, rational 'man' [sic] who, through positivistic science and technology, has sought to control Nature and constitute a totalizing and universal 'Truth.'" (p. 3).

Fernández-Balboa (chapter 1) traces the roots of modernism to the intellectual movement of the Enlightenment and goes on to list a number of terms associated with the Enlightenment or modernism: positivistic, analytical, scientific, mechanistic, anti-metaphysical, sensationalist, associationist, empiricist, social contract theory, individualism, natural right theory, pursuit of self-interest, dualism, and capitalism. In the space of three paragraphs, he cannot provide explanation of each of these terms. The reader, unless previously well-informed about modernism, is likely to be overwhelmed by the terminology and left without a clear understanding of the concept of modernity.

The authors of the subsequent chapters seem to presume that the reader is familiar with modernity. They provide additional clues to its meaning but no systematic explanation. Brustad (chapter 6) describes the dominance of positivism in physical education research and the general acceptance of the principles of reductionism, objectivity, and quantification. Perhaps the clearest picture of modernity and its dominant influence in physical education is provided by Tinning (chapter 7). His stories and examples from professional practice move beyond the list of characteristics and provide a more accessible understanding of the concept. Ironically, he achieves this clarity by creating an either/or category system (performance and participation) that exemplifies the dichotomous thinking of modernism.

Despite the sketchy presentation of modernism, the book makes the case that the profession of human movement is deeply rooted in this intellectual tradition. Most of us would concur with this conclusion. Our personal experience is likely to confirm the widespread emphasis upon science within human movement programs. However, for many readers, the positivist scientific tradition is likely to be viewed, not as a problem, but as the natural and appropriate way to approach the field.

The authors' goal is to encourage the readers to critically examine this taken-for-granted perspective grounded in the assumptions of modernity and to reject it. In chapter 1, Fernández-Balboa refers to a grave global crisis as evidence that the promises on which modern civilization is based have failed. He states that, besides politicians and corporate tycoons, the rest of the people

> ... have sadly sunk into lower and lower levels of hopelessness in a world where crime, war, hunger, poverty, pandemics, environmental destruction, etc., have been the norm. . . . These are clear signs of an acute global crisis, a crisis brought about by modern thought and action. (p. 4)

Persuading others to accept his conclusion does not seem as easy as he implies. The challenge is to convince the readers that the problems cited are evidence that modernity has failed. The socially dominant view is that the principles of modernity are valid but that our efforts to implement them needs to be improved. At various points throughout the book, the authors acknowledge that challenging the dominant position will not be easy. The hegemonic process by which people are persuaded to accept the legitimacy of the dominant system of beliefs is described briefly in chapter 1. The enormous power of resistance to change is alluded to throughout the following chapters.

Despite hegemony and resistance to change, Fernández-Balboa suggests that a transition to a new era of postmodernism has begun. He implies that the shift to postmodernism is inevitable and that this new era will bring change in every aspect of life, including the profession of human movement. However, the explanation of post-modernism is even more sketchy than that of modernism. There is a suggestion that personal identities, cultural meanings, and social relations are socially constructed but what this means is never really made clear. The result is that the reader is left with the feeling of being asked to endorse something not clearly understood, a slogan not a theoretical perspective. The risk is that this gives an impression that the book is polemical rather than persuasive.

Earlier work that has adopted a "radical" perspective has been criticized by others in the field who list their concerns with the literature as follows (a) polemical presentations and vitriolic language, (b) zero-sum arguments that polarize, (c) cartooning, (d) a lack of evidence to defend assertions, and (e) assumptions that it alone occupies high moral ground (O'Sullivan, Siedentop, and Locke 1992). To counter these criticisms, authors need to provide clear explanations and thorough analyses that acknowledge the complexity of these issues.

The problem is that modernism and postmodernism are complex philosophical positions that cannot be easily explained in chapter 1. This is particularly the case with postmodernism, which is difficult to comprehend even when discussed more fully. Although reading the book as a total text may extend the reader's understanding of these concepts, this may not be readily achieved. Brodribb (1992), in discussing the meaning of poststructuralism and postmodernism, states that, "Profoundly elusive, purposively ambiguous, these are terms which are not used systematically, and about which there is not consensus" (p. 8). The lack of clarity about postmodernism is compounded by the authors' tendency to confound postmodernism and critical theory. The two terms are not synonymous but the distinction between the two is not clearly drawn. More importantly, the tension between postmodern theory and critical theory is not discussed in depth.

Postmodernism asserts that identities, meanings, and relations are never fixed and constant but always being created and recreated. This lack of certainty and rejection of grand narratives has created problems for critical theorists and feminists. The commitment of critical theorists and feminists is to empower those who have been oppressed by the current social-political-economic system; their aim is to struggle more effectively to create new cultural and social patterns that are more just and liberating. They confront the challenge to identify the basis for the social vision that guides the transformative effort. At stake here are some important issues: What kind of a world do we want to create? What is our rationale or justification for our vision? Most of us were raised to take for granted the assumptions of the modern era, a belief in human progress through rationality and objective science. This was the *metanarrative*, the grand scheme that provided the foundation for our belief that a prosperous society could be created by rational action.

This taken-for-granted view of the world has been shaken by postmodernism which views all knowledge as socially constructed, partial and incomplete, and context-specific. Postmodernists note that our assumption of a universal, objective standard of truth tends to legitimate the truth of those with power and dismiss alternative views of truth and knowledge. Postmodernists *reject* not only the metanarrative of modernism, but also *the very possibility of meta-narratives*, the possibility of generalizable and universal claims of any sort. One result of this critique of mainstream thought has been to heighten anxiety—fears that if we have no objective foundations for truth, we will "fall into the abyss of relativism, skepticism, and nihilism" (Stanley 1992, p. 151).

The problem for critical theorists and feminists committed to change remains: If we take seriously the arguments of the postmodernists, how do we ground our efforts for social transformation? Can we create a more just society without simply substituting a new metanarrative for the old one? Can we describe oppression and identify alternatives without slipping into essentialism; that is, without assuming that all men or all women (or all members of any group) have the same fundamental nature? Can we do more than criticize the status quo? In addition to our language of critique, can we find what Giroux (1988a, p. 135) has called a "language of possibility?" Can we develop self-conscious and self-critical practices that create change?

Some are concerned that the uncertainty inherent in postmodernism will paralyze efforts for social change, that the absence of foundations for defining social values is a form of neoconservatism that will undermine change (Harding 1990). Brodribb (1992) goes further in her critique, viewing postmodernism as inherently misogynist. Others have attempted to reconcile feminism and postmodernism, suggesting that the solution is to ground our efforts on an understanding of difference, accepting that knowledge is always provisional, open-ended, and relational. The challenge for transformative postmodern theory is to create a process that opens new possibilities without creating a new grand scheme, a new world order. Hartsock (1990) describes it this way:

We need to develop our understanding of difference by creating a situation in which hitherto marginalized groups can name themselves,
speak for themselves, and participate in defining the terms of interaction, a situation in which we can construct an understanding of the
world that is sensitive to difference. (p. 158)

Often, discussions of difference (including those of some critical
theorists) assume that individuals have a known, single identity
that fixes them in a particular place in the social system. Postmodernism denies the unity and stability of identity. Individuals occupy
multiple standpoints at any one time and different standpoints at
different times. Because the individual is not a single, stable subject,
self-knowledge is always incomplete, as is the ability to be completely open with others. Within the individual, there are layers of
meaning that the person struggles to comprehend and express (von
der Lippe, chapter 3). Orner (1992) suggests that this struggle for
identity has broader social significance:

Feminist poststructuralist discourse views the struggle over identity
within the subject as inseparable from the struggle over the meanings of identities and subject positions within the culture at large.
(p. 74)

The connectedness of the personal struggle for identity with the
social struggle over positions in the culture is of particular relevance
for the profession of human movement because of the significance of
embodied knowledge in those struggles. Embodied knowledge—
ways of speaking and moving, ways of using and caring for and presenting our bodies—becomes a cultural language that serves to
create and convey our identities. The challenge that confronts the
profession of human movement is first, to acknowledge the central
role of embodied knowledge in the field, then to decide what postmodernism means for fields that specialize in embodied knowledge.
As suggested throughout this book, postmodernism requires that we
stop assuming that there is a single, universal truth about what it
means to have a healthy body, an attractive body, a skilled body.
Those terms have contested meanings and have to be struggled for
and defined by different people in different contexts. The practical
question for the profession is how to create educational environments that empower individuals to create meaningful lives and a
just society.

A Practical Analysis

What are the implications of transformative postmodern theory for professional practice in human movement? The emphasis throughout the book is on engaging professionals and students in critical reflection about ideas, assumptions, and experiences that might have been previously seen as unproblematic. The authors extend the analysis to a number of specific topics and, to varying degrees, discuss the implications for professional practice.

One area of concern is the nature of school physical education in the postmodern era. Kirk (chapter 4) suggests that to survive in the postmodern era, school physical education programs need to be culturally relevant. While several of the chapters focus on the sociopolitical aspects of postmodernism, Kirk (chapter 4), Fahlberg and Fahlberg (chapter 5), and Schwager (chapter 9) speak most directly to the personal meaning dimensions of the postmodern cultural transition. Kirk's discussion of the social construction of bodies and Fahlberg and Fahlberg's emphasis on the relationship of movement and health to human freedom problematize traditional assumptions of physical educators. They also highlight that students in the postmodern era bring expectations and perspectives of the world that differ greatly from those of their teachers. Kirk argues that the cultural changes of "high modernity" have made traditional physical education obsolete.

To achieve cultural relevance, Kirk recommends that students elect from a wide range of activities and contract to complete required number of hours of participation in or out of school. But to differentiate this educational program from a merely recreational program, students need to be engaged in critique of aspects of the "contemporary popular physical culture which present idealised and glamorised body images" (p. 59). Fahlberg and Fahlberg (chapter 5) suggest using critical reflection about movement experiences to enhance personal development and to enable the individual to define for oneself what is healthy and what is not. Schwager (chapter 9) suggests that teachers must prepare students to be morally responsible citizens by modeling socially just behaviors and engaging students in dialogue about justice. The common theme of each of these chapters is the need to engage students in critical reflection and dialogue. Furthermore, Brustad (chapter 6) suggests that, to be relevant to the postmodern era, scholars need to change what they study and how they study it.

Several authors recognize that changing school physical education will require major changes in university programs, especially in teacher education. Ingham (chapter 10) proposes a radical restructuring of departments of what he calls "physical cultural studies" in which "exercise practices, health practices, nutritional practices, sport practices, leisure practices, and pedagogical practices [are not] hierarchically arranged, categorically boxed, discrete academic territories" (p. 171). Fernández-Balboa (chapter 8) proposes redesigning PETE programs to engage teacher educators and prospective teachers in "critical reflection and analysis with the intention to uncover the dominant ideologies; deconstruct taken-for-granted knowledge, meaning, and values; and practice pedagogy as a means for human agency and civic and environmental responsibility" (p. 127). He provides examples of curricular themes and pedagogical practices intended to support this aim. Those teacher educators who wish to implement critical reflection in pedagogy courses could use his ideas as a starting point.

The question for teachers and university faculty is how to go about involving students in critical reflection. A recent study of feminist physical education teachers in Australia indicates that the dominant technocratic culture in school physical education makes it difficult to engage students in critical reflection (Smeal, Carpenter, and Tait 1994). One source of help may be Hellison's writings related to the use of reflection to enhance the personal and social development of student (Hellison 1978, 1985, 1995; Hellison and Templin 1991). Although Hellison does not take a critical theoretical stance and does not directly address issues such as gender, race, and class, his work aiming to empower at-risk youth provides extensive examples of practice and may be useful to those interested in transformative pedagogy.

Preparing physical educators to be transformative teachers will require more than incorporating critical reflection in pedagogy courses or social science courses or increasing the number of such courses. Tinning (chapter 7) cites evidence of student resistance and marginalization of the social sciences. Similar resistance has been reported by Dewar (1986, 1991) and Gore (1990, 1992). Schwager (chapter 9) describes the difficulties she encountered trying to get university students to focus on moral responsibilities and social justice. Student resistance is not unique to kinesiology and physical education (Ellsworth 1989; hooks 1989). Much of this resistance is based on university students' concern about acquiring the technical skills they need to succeed in teaching or in other aspects of the human movement profession. The challenge for those committed to transformation and critical theory is to build a curriculum that includes critical reflection as well as professional skills.

Students are not the only source of resistance to social change. Both Fernández-Balboa (chapter 8) and von der Lippe (chapter 3) refer to professional risks for faculty who challenge the status quo. The implication is that transformative pedagogy requires an overt political stance professionally and personally. And such a stance requires courage.

A Personal Analysis

The transformative postmodern perspective requires that one bring a "critical awareness" to one's daily life. That awareness encourages one to examine how the meaning of an event is being constructed by various people and the effect of those constructed meanings on relationships and power arrangements. Critical awareness also leads one to examine personal reactions to the event and to recognize how those reactions relate to elements of personal experience and identity. Let me provide an example.

During the time I was working on this chapter, I attended the LPGA Nabisco Dinah Shore Golf Tournament in Palm Springs, California. I would like to use this event as an illustration for the personal relevance of the transformative postmodern perspective. The Nabisco Dinah Shore is the first major tournament of the season for the LPGA (Ladies Professional Golf Association). The tournament began in 1972 and has been sponsored by Nabisco since 1981. The tournament draws a large field including most of the top women golfers and has been won by many of the women in the LPGA Hall of Fame. Major corporate sponsorship and its location in Palm Springs give the tournament a high profile. The tournament is preceded by a two-day Celebrity Pro-Am event that draws many celebrities: the program lists former President Gerald Ford, Bob Hope, Robert Wagner, Johnny Bench, and Joe DiMaggio among past participants. The slick sixty-four-page magazine program includes articles about top golfers and the glamour of the event as well as thirty-nine pages of advertising. Because of its location in a premier resort community and its proximity to Los Angeles, the tournament draws large galleries. It is also nationally telecast.

The Nabisco Dinah Shore Golf Tournament is also a major event in the lesbian world. The *Lesbian News*, a monthly magazine published in Los Angeles with a circulation of 100,000, carries extensive advertising and publishes a special supplement listing thirty different lesbian social events held in Palm Springs during Dinah Shore weekend. Thousands of west coast women travel to Palm Springs for the festivities, filling most of the major hotels. Many attend the golf tournament; others come only for the social events.

In March 1995, I attended the Dinah Shore tournament for the first time. My partner and I drove to Palm Springs for the day on Saturday, March 25. It was a beautiful southern California day, crystal clear, with temperatures in the low 70s. About two hours east of Los Angeles, we exited the freeway on Bob Hope Drive and drove to the course at Mission Hills Country Club. Beyond the lush green of the golf course, we could see snow in the mountains above the town. We decided to walk the course and watch different groups of golfers rather than follow the leaders. To our eyes as amateur golfers, all of the players demonstrated a grace and skill that we admired. We watched the golf, visited with friends, and enjoyed the beauty of the place. After the tournament, we stopped by a "Lesbo-Expo" where fifty vendors and exhibitors displayed and sold books, crafts, T-shirts, and other wares. It was a delightful way to spend a spring day.

The next day we watched the final round on television. The broadcasting team, led by Brent Musberger and Judy Rankin, focused their attention on what they referred to as the "dream team" final threesome of Laura Davies, Tammie Green, and Nancy López. Green had been leading at the beginning of the final day but López played well on the front nine and seemed the favorite as the final holes neared. Yet, on the last few holes, López faltered, hitting the ball into the water on the 18th. Much to the surprise of the television crew, Nanci Bowen, a relative unknown in the next to last group, won despite a bogey on the 18th.

The outcome of the golf tournament provided unexpected drama, but the aspect of the television report that struck us was the heterosexist emphasis of the coverage. The cameras showed very few shots of the crowd and none of the identifiable lesbian presence. The reporters repeatedly mentioned the husband and children of Nancy López and the husbands of other married golfers. They mentioned celebrities in attendance but not the presence of Colonel Grete Cammemeyer, the lesbian whose battle to stay in the military had recently been the subject of a television movie.

The heterosexual tone of the presentation of the golf tournament was undoubtedly influenced by the fiscal realities of the professional golf tour. Commercial sponsors provide prize money, and advertisements pay for the television coverage. Product sponsorships are also an important source of income for individual golfers. Professional athletes are aware that, despite her status as one of the

top athletes in the world, Martina Navratilova has attracted few
commercial endorsements since her public declaration that she is a
lesbian. In order to sustain commercial support for the LPGA, tournament officials, golfers, and television networks collaborate in presenting a picture of women's professional sports that denies the
presence of lesbians and reinforces traditional heterosexual values.
The net effect was to make lesbians invisible at an event that is one
of the major lesbian festivals in the nation. And as members of the
viewing audience, it made us feel invisible and powerless as well.[1]

I am struck every day by the contrast of the high visibility of my
professional life and the invisibility of my personal life. Throughout
my career, I have chosen what I have called the "glass closet." I live
my life in a lesbian world that is relatively apparent to those around
me, but I have not proclaimed my lesbian life-style at work. Others
can see into my glass closet but, because I have not named it, they
can ignore my sexual orientation if they so choose. I am uncertain of
the professional impact of this path. It may have enabled me to move
into administrative roles that would have otherwise been blocked (or
it may have made no difference). But I know that the personal impact has been a disjointed sense of self and a feeling of being invisible as a person in my professional world.

Dividing my personal and professional worlds has become more
of a problem as I have moved into higher administrative positions.
Deans and vice-presidents are expected to attend many social events
as part of their professional roles. I am less able and less willing to
maintain the charade of being a single woman when I am not. Recently, I interviewed for a position as vice-president; the president of
the institution made inquiries about my personal life and eventually
did not hire me. My "glass closet" failed to protect me, and I was
frustrated and angry and ready to break out. I subsequently was selected as Provost and Vice-President for Academic Affairs at San
José State University, one of twenty-two campuses of the California
State University which has a policy forbidding discrimination on the
basis of sexual orientation. Despite that policy, a number of friends
and colleagues have cautioned me that publicly identifying myself as
a lesbian might cause problems in my role as Provost. But it is time
for my act of courage.

The effect of my decision will be political but ultimately the motivation is personal as well as political. I need to reintegrate my personal and professional identities for my own well-being. But as
Orner (1992) points out, the struggle over personal identity cannot
be separated from the struggle for position in the culture. Just as my

silence had an impact, the act of naming my identity and describing my experience will affect the social-political reality of those with whom I interact. My regret is not having spoken sooner, but my past actions were grounded in my experiences and situation. Postmodernism and critical theory provide a useful perspective for understanding the circumstances that have shaped my decisions and the personal and political consequences of my actions. Feminism has provided a lens to link my experience and values to issues related to the rights and well-being of all women. The understanding that emerges from such critical analyses has given me the courage to act.

Closing Remarks

Postmodern theory postulates that each of us has multiple identities, different standpoints in different contexts and at different times in our lives. Critical reflection enables us to examine how personal meanings and identities relate to social-political-economic relationships and issues. The authors of this book have presented a series of analyses intended to challenge the taken-for-granted world view of readers. They have discussed the implications of their analyses for professional practice in human movement.

The book is intended to stimulate reflection and dialogue. Estimating the impact that the book will have is difficult. Despite their criticisms of the "radical" literature in physical education, O'Sullivan, Siedentop, and Locke (1992) acknowledge the need to build a critical tradition in the field and credit the "radicals" with creating serious consideration of moral and social issues. This book should be viewed as an important contribution to that discourse. However, as Fernández-Balboa (chapter 8) and Hellison (1992, and also chapter 12) point out, the final step is transformative action. Educators often avoid political activism, justifying their "neutrality" on the basis of their responsibility for protecting objectivity and free speech. In the name of protecting others' right to speak, they give up their own voices. Certainly we have a moral obligation to respect the rights of others. But we also have a moral responsibility to speak and act in ways that reflect our own values. The form that our actions take will differ, but the challenge this book poses is for each of us to be fully engaged in *reflection* and *action*.

CHAPTER 12

A Practical Inquiry into the
Critical-Postmodernist Perspective
in Physical Education

Don Hellison

Introduction

The critical-postmodernist perspective (CPMP)[1] is well represented in this book. Every chapter argues in one way or another for
the dawning of a postmodern era that requires a socially critical
physical education. Scholars who have distinguished themselves in
critical and postmodernist theory such as George Sage, Alan Ingham,
Gerd von der Lippe, David Kirk, and Richard Tinning are joined by
other writers with fast-growing reputations in critical-postmodernist
thinking, for example, Juan-Miguel Fernández-Balboa and Larry
and Lauri Fahlberg, as well as scholars such as Bob Brustad and Sue
Schwager who have developed reputations in their respective sub-
disciplines and now offer a more critical-postmodern perspective.

In his introduction, Fernández-Balboa (chapter 1) states that
the intention of this book is "to delineate ideological and political
markers to enable readers to see the [human movement] profession,
not in a vacuum, but as a political movement of sorts; a diverse collection of communities; and a forum for acceptance, equality, and
freedom" (p. 7). I believe that the chapters in part I of this book do a
very thorough job of delineating the CPMP's ideological and political
implications; yet I believe that in order to improve human movement
practice, *other alternatives need to be reflectively explored as well*
(although this is not necessarily the task of these authors). The authors also offer some new maps and new borders for putting the critical perspective into practice in professional preparation and sport,

197

exercise, and physical and health education programs. However, with a few exceptions, most of the chapters fall short of offering specific examples of the CPMP in practice, especially in sport, exercise, and physical and health education programs. Perhaps I have misinterpreted the intention of this book, or perhaps the barriers described by Sage, Ingham, Kirk, and others prevent more applications of the CPMP. But it is my strong belief that if critical-postmodernist scholars really want to impact professional preparation and sport, exercise, and physical education programs, their activism needs to be extended into the world of practice. I will borrow some ideas from practical inquiry to elaborate on both of these points.

Practical Inquiry

Schubert's (1986) comparative analysis of paradigms for curriculum inquiry contrasts the critical perspective not only with the empirical analytic paradigm, which Brustad does in this book (and which also surfaces elsewhere as well), but with "practical inquiry" (see Martinek and Schempp 1988; and Richardson 1994, for additional clarification of, and support for, practical inquiry).

Practical inquiry can provide the kind of activism needed to transform ideas into practice, an aspect that is mostly missing in these chapters. Schubert shows that, unlike both the empirical-analytical and critical perspectives, practical inquiry requires interaction with practice in a specific setting. Critical-postmodernists may make this claim also, but they seem to view practice at a distance rather than in continual interaction with a specific setting. The result of practical inquiry is *"situational* insight and understanding" (italics mine) leading to an "increased capacity to act morally and effectively" (Schubert 1986, p. 289).

Practical inquiry, with its roots in the work of John Dewey, William James, and Joseph Schwab, has company in the form of other emerging alternative research methodologies such as action research (Stanley 1995), teacher-as-researcher (Duckworth 1986), reflective practice (Schön 1991), and curriculum-as-craft (Kirk 1993). Taken together, these approaches, which (with a bit of license) I will refer to as practical inquiry, argue for wading in the swamp of practice (Schön 1987); setting the problem (Lawson 1993b, 1984); creating, borrowing, or modifying a set of ideas to address the problem; putting the ideas into practice; shuttling back and forth between practice and reflection; doing an ongoing evaluation; and sharing the results of this situation-specific process so that readers/listeners can judge the merits of the ideas for themselves and, if they so desire, put them into practice in their own setting. Similar ideas with regard to this process have been advocated by Paulo Freire under what he calls "praxis" (Freire 1970).

For me, practical inquiry, as I have defined it, represents an important but mostly missing dimension of the CPMP. Practical inquiry provides a way to analyze the CPMP's impact on sport, exercise, and physical and health education practices and has the added benefit (beyond exploring the theory-practice gap) of suggesting that the two perspectives might complement each other.

Signs of the Practical in the Critical-Postmodernist Perspective

The chapters in part I do contain some interest in practical application. Kirk (chapter 4) makes recommendations for changing the practice of physical education (e.g., broadening the kinds of activities offered, developing individual contracts that can be fulfilled outside school, emphasizing critical reflection in PETE programs, and encouraging more collaboration among school and community physical educators). More importantly from a practical inquiry perspective, Kirk includes some examples of change in some Australian secondary schools such as inclusion of "critical analysis of the social construction of gendered bodies in and through sport and exercise [and] learning outcomes which problematise compulsory heterosexuality and hegemonic masculinity" (p. 59). While Tinning (chapter 7) only emphasizes *advocacy* for making physical education more than a rhetorical site of social change, particularly with regards to issues of equity and injustice; he has put these ideas into practice in a curriculum development project in the Victorian secondary schools (Tinning and Fitzclarence 1992). This project is an excellent example of practical application of the critical perspective, although it is not cited in Tinning's chapter.

Schwager's arguments (chapter 9) on critical moral issues focus on the transformative function of schools. Her recommendations for changing practice are a bit eclectic, for example, classroom management which is a dominant modernist practice, Mosston's spectrum of teaching styles which has been around for well over a decade, and my approach to teaching responsibility. Her emphasis on social justice and empowerment is clearly grounded in the critical perspective, but the somewhat eclectic nature of her recommendations suggests that she is too close to practice to fully embrace the CPMP. My explanation (not necessarily hers) is that she works directly with teachers and schools on a regular basis and, in attempting to deal with the reality of schools, teachers, and students, she cannot afford to adopt a "pure" CPMP. This is precisely what happens when one conducts practical inquiry and underscores the tension between practical inquiry and the CPMP.

Fernández-Balboa (chapter 8) calls for civic and political activism among physical education teacher education (PETE) students: "They must learn how 'the physical' is constantly used to establish unequal relations of power, and how sport and physical education serve to perpetuate the dominance and privilege of certain elite groups [so that PETE students may be able] to struggle against them" (pp. 127–128). To do this, he provides a list of reflection questions and some methodological practices to be used in the teacher education classroom. He also argues for students getting involved in political action by "writing letters and articles . . . , organizing peaceful demonstrations . . . , and utilizing diverse physical and artistic forms to foster equity and environmental conservation" (pp. 136–137). From a practical inquiry perspective, one of the strengths of Fernández-Balboa's arguments (chapter 8) is his concrete ideas about how to do critical pedagogy in teacher education as well as his own experiences in carrying out these ideas in his teacher education practice (Fernández-Balboa 1995). However, from his chapter, I did not get the idea that *teaching kids differently* was central to PETE students becoming "agents of change." My attraction to practical inquiry is precisely for this reason. Yes, we need to be "models" in our university classrooms, but in my view we also need to witness our beliefs by "wading" alone or, better yet, with our PETE students in the "swamp of practice" (as Schön puts it), exploring alternative ways of working with kids.

The point is that critical alternatives need to be tested in practice. That means going out and working with kids, or else working in collaboration with practitioners who share these values, and then putting these ideas into practice. It seems to me that if we want to establish a genuine dialogue with practitioners, rather than just with each other in academe, we need to demonstrate that we, too, have struggled with the realities of practice and have something concrete to share. This may be a key advantage of practical inquiry—it forces us out of our comfort zones of publishing and speaking to academic audiences and into real world trials.

Purpose and Values

According to Schubert (1986), the critical perspective aims at "emancipation, equity, and social justice" through empowerment (p. 315). Most of these concepts appear in all of the chapters and provide a value-based purpose for the practice of specific sport, exercise, and physical and health education programs. These concepts do take on a variety of forms and interpretations because, as these chapters point out, there is no single, unified CPMP. For example, for Fernández-Balboa, human freedom and justice are paramount values, but he emphasizes not only "a more equitable society," but also preservation of the natural environment as well. Fahlberg and Fahlberg (chapter 5) pay some attention to psychological repression as one of "the two bars of imprisonment" (the other bar being social oppressive forces). Tinning (chapter 7) expresses concern about the "problematic practices" of the "movement culture," especially with regard to their dominant performance orientation and the elitism that results from such an orientation. However, he falls short of viewing physical education as a vehicle for social change outside gyms and playing fields, calling the idea of physical education correcting "the massive, social and economic problems of the world . . . ridiculous"[2] (p. 105).

Throughout the book, there is much agreement as well. As one example, Tinning opposes the human movement professionals' penchant for setting "out to regulate the lives of people in their own best interests [with] a mix of paternalism, evangelicalism, and arrogance" (pp. 106–107), a point also made by Fahlberg and Fahlberg (chapter 5) in their criticism of the prescription-compliance oriented medical model, and by Kirk (chapter 4) and Fernández-Balboa (chapter 8).

Practical inquiry is not as concerned with these theoretical, though important, issues as it is with finding solutions to specific problems. For practical inquiry, the "source of problems is found in a state of affairs, not in the abstract conjuring of researchers who tend to imagine similarities among situations that cannot be grouped together defensibly" (Schubert 1986, p. 289). The focus is on situational problem solving. However, as Lawson (1984) reminds us, this process of framing and naming problems is never value-free, nor is the practical inquirer's selection, modification, or creation of theories to address these problems.

Therefore, practical inquiry is open to a variety of value orientations. What can be said for practical inquiry on this point is that it is based on the vague humanitarian value of improving professional practice in order to serve people better. This leaves the door open to a variety of purposes, including Fernández-Balboa's three versions of pedagogy—technical, humanistic, and critical. The technical purpose finds its value basis in the application of research drawn primarily from the empirical-analytic perspective which Brustad (chapter 6) and Ingham (chapter 10), among others, find problematic. The humanistic purpose, which emphasizes self-esteem, self-direction, cooperation, and so on, is also problematic, in this case because, according to Fernández-Balboa (chapter 8), "the learners' education is confined to personal levels; the emphasis on social transformation is usually left out" (p. 127). These criticisms suggest that the CPMP is a more appropriate value orientation for carrying out practical inquiry (Kemmis and McTaggart 1988; Stanley 1995).

The gap in purpose between the technical and critical perspectives is greater than the difference between the humanistic and critical perspectives. However, the humanistic/critical difference is not insignificant. In Ennis's value-orientation research, urban physical education teachers clearly preferred a more humanistic orientation (Ennis 1994) to one emphasizing social reconstruction (Ennis, Ross, and Chen 1992). Further evidence for this difference can be found in analyses of my work by Bain and Jewett (1987) and Shields and Bredemeier (1995), who find problematic its failure to move beyond empowerment of the individual.[3] While practical inquiry may remind critical-postmodernists of the need to put their ideas into practice, as well as a way for doing so, the CPMP reminds practical inquirers of the need for a clear sense of purpose and of the importance of problem setting.

Research Perspectives

Brustad (chapter 6) criticizes the dominant empirical-analytic research paradigm for, among other things, conducting research nomothetically (i.e., comparing groups and reducing findings to generalizations) rather than idiographically (focusing on individual differences), for ignoring the social context, and for lacking commitment to changing and improving social structures. The CPMP certainly pays attention to the social context and is highly committed to improving social structures. Whether it honors the idiographic seems problematic. Although the CPMP does not share the nomothetic assumptions of empirical-analytical research (e.g., objectivity, quantification, and reductionism), much of its rhetoric focuses on oppressed groups (e.g., "minorities") rather than on individuals. Fahlberg and Fahlberg (chapter 5) pay more attention to "the individual" than other authors, but they place this interest within a nomothetic human development context. Practical inquiry, on the other hand, focuses on a specific social setting and is very careful with generalizing results and recommending policy, to the point of eschewing anything resembling laws or principles, especially when referring to large groups and contexts.

Moreover, Ingham (chapter 10) suggests critical self-reflexion as a viable way to evaluate the dominant research perspective in universities. In his analysis, he criticizes the "technocratic intelligentsia" who populate our sub-disciplinary structures (i.e., "our ghettos of knowledge"), and calls for more attention to the conceptual and cross-disciplinary orientation of the "humanistic intellectual," primarily by studying practices in physical culture and focusing more broadly on the body and its socially constructed meanings (a point shared by Kirk, chapter 4). This means, in part, not "privileging" those who do empirical-analytic research over those who use other paradigms. Ingham also criticizes studies that focus on internal validity rather than on external validity, because in "the swamp of practice" the realities and the rights of those studied are usually traded off for the safety of a controlled environment. He states that, if we want to "humanize our public interventions," we can't "dehumanize our research."

In their respective chapters, both Brustad (chapter 6) and Tinning (chapter 7) also criticize our current conception of research arguing, as Ingham does, for a more humanistic research agenda (e.g., human development instead of human performance, making a contribution to bettering the world) and for broader, less specialized thinking in our research endeavors.

Although not well recognized as a human movement research methodology, practical inquiry meets some of the criteria described above and, therefore, offers an alternative to the empirical-analytic paradigm (Georgiadis 1992). For example, it requires less specialization, as Ingham and Tinning advocate, because messy real-world social and educational problems do not fit neatly in anyone's specialization and, therefore, are not amenable to "treatment" by specialized forms of knowledge. Practical inquiry is also consistent with Ingham's emphasis on external validity and working in a natural, uncontrolled environment. Brustad's and Tinning's argument for a type of research that serves humanity opens the door to the utilization of a practical inquiry that focuses on human problems such as

those found in sport and physical education practices. The scope of
practical inquiry is very wide. For example, in addition to a value-
based sense of purpose that embraces human development with regards to the "underserved" (e.g., the poor, "minorities"), ideas well
represented by the critical-postmodernist scholars, an emphasis on
elite performers might also be easily embraced by those practical inquirers who might view improvement of elite athletes as a human
problem.

Closing Remarks: Integrating the Two Perspectives

Besides myself, others have also questioned the lack of practical
activism in the critical-postmodernist ranks (e.g., Lawson 1993b;
O'Sullivan, Siedentop, and Locke 1993). In the same vein, Prain and
Hickey (1995) attempt to redress the CPMP's failure "to inform and
influence practice" (p. 77) by utilizing discourse analyses of teacher-
student verbal interactions to redefine effective teaching and to
explore the "link between intention, implementation, and overt
learning" (p. 87).

In its advocacy of a specific set of values, the CPMP challenges
practical inquiry and, in so doing, underscores the importance of
having a clear sense of purpose and of asking the question, What is
worthwhile to know and experience? (Schubert 1986). Practical inquiry, on the other hand, reminds critical-postmodernists of their
failure to practice what they preach, especially in sport, exercise,
and physical and health education settings. As I have pointed out,
some scholars have began to bring the CPMP and the practical inquiry perspective closer together (Stanley 1995).

I have attempted to integrate some aspects of practical inquiry
and critical-postmodernist thought in the courses I teach. The practical inquiry process of problem setting, identifying a set of ideas to
address the problem, putting these ideas into practice over and over
again, conducting an ongoing evaluation, and sharing these ideas
and experiences is evident in my work with kids (Hellison 1995). Yet,
I acknowledge that this process is very difficult to carry out in traditional classroom-gymnasium settings in professional preparation
courses. Therefore, every semester I offer a nontraditional professional preparation elective course available to students of different
disciplines (e.g., PETE, kinesiology, social work, psychology, education) who are attracted by my humanistic beliefs and values and
wish to work with inner city children and youth.[4] The course (i.e.,

"At-risk Youth Leadership") reflects my own sense of purpose as well as my belief in practical inquiry and is designed to analyze and seek solutions to specific educational problems with inner-city kids. It consists of two parts: (a) an internship in one of my "Responsibility in the Gym" programs for inner-city kids in which interns are required to assist me in putting my ideas into practice; and (b) a weekly seminar to plan and evaluate specific lessons, to problem solve specific incidents, and to analyze the theory and practice of this approach by discussing reading assignments as well as reflecting on their internship experiences. Out of these discussions, new ideas and strategies are forged, tried out, and critiqued. Through this course, students experience the processes of practical inquiry and critical reflection as well as an alternative PE perspective.

My undergraduate PE curriculum course required for PETE students, however, takes a different approach. I can only raise a few of the questions relevant to practical inquiry, because this perspective is difficult to apply in a traditional professional PETE structure.[5] So I borrow a page (not the whole book) from the CPMP by taking a reflective approach that includes autobiographical reflection, reflection on current and emerging practices, and reflection on dominant societal practices (Hellison and Templin 1991). The students' final project is their own PE curriculum model; their grade is not based on conformity with my values, but on their development of a defensible, coherent model that shows evidence of the critical reflection process. High marks have been earned for multiactivity models, sport education models, even models with a heavy dose of competitive achievement. In this course, empowerment does not mean guided discovery toward predetermined goals but, rather, wide open reflection with a few negotiable guidelines of acceptable practice.

For me, the CPMP is not the only way to do practical inquiry. Nevertheless, I recognize that the CPMP can add value to practical inquiry for it calls our field's attention to important issues of social justice and empowerment and reminds us that our purpose must be guided by what we consider to be worthwhile to know and experience. This is a crucial question, and we need to ask it again and again in our research and in our practice. Practical inquiry, on the other hand, reminds critical-postmodernists to extend their thinking into practice by going there, not as visitors but as activists. Only then, it seems to me, can their case be truly advanced.

CHAPTER 13

Defining the Dreaded Curriculum:
Tensions Between the Modern
and the Postmodern

Catherine D. Ennis

Introduction

A curriculum is the product of the beliefs and values of the individuals who design and implement it. It may facilitate or constrain an individual's educational experiences and aspirations. Like other educational constructs, curricula are value based. Although they reflect the dispositions or ideals of their advocates, they also embody the "nightmares" or worst case scenarios of those who oppose those ideals and dispositions. In other words, most sets of valued propositions esteemed by one group also are "dreaded," disdained, and often disregarded by other groups. In this chapter, the term, "dreaded curriculum" refers to the programmatic image that is most ridiculed and, at times, feared by a particular group or individual. I will argue that often one person's valued ideal is another's dreaded curriculum.

The dreaded curriculum reflects the philosophical positions that we hope are *not* incorporated into our educational programs. I have borrowed the term from work by Oyserman and Markus (1990) associated with the psychological concepts of ideal and dreaded self. If we believe that curricula are derived from our personal and professional beliefs, then the metaphorical use of these psychological concepts may be especially fitting. Oyserman and Markus explain that, unlike the ideal self that motivates one to achieve through the promise of success, the dreaded self motivates through avoidance of fear. In the same vein, Khmeikov, Makogon, and Power (1995) suggest that the dreaded self has significance as a moral motivator.

207

Individuals attempt to elude the dreaded self by engaging in moral acts, thus avoiding negative responses such as harming others. Khmeikov et al., describe this phenomena as part of a moral code that can be used to protect our self respect from shame and guilt (Fahlberg and Fahlberg, chapter 5). This is considered different from promoting self-respect by pursuing positive courses of action that empower the ideal self. When moral concerns are most salient, psychological and rhetorical definitions of the dreaded self focus on avoidance and fear (Khmeikov et al. 1995).

A dreaded curriculum, then, is one that individuals wish to avoid because it is offensive or threatening to their moral code. Upholding this moral code requires rejecting this curriculum and embracing one that is self-protective and does not damage the tenuous balance between educational and moral responsibility. Some authors of the previous chapters (e.g., Fahlberg and Fahlberg, chapter 5; Ingham, chapter 10) have provided a vivid picture of the dreaded curriculum of rationality. But because of their own philosophical biases, they have not provided a critique of their own perspective. Therefore, in the first section of this chapter, I will present a critique of the dreaded emancipatory curriculum from the perspective of their adversary—rationality. In the second section, I will adopt a less adversarial position and propose a stance on these issues based on "reasonableness."

The Postmodern Perspective on Rationality

The authors of the chapters in part I have taken great care to develop and present their ideas of dreaded and ideal curricula from a socially critical, postmodern perspective. Usually, in the opening section of their respective chapter, they have chosen to present rationality as the dreaded curriculum. Almost without exception, they have conceptualized it as the curriculum or programmatic agenda currently in operation in schools and universities. For example, Ingham describes the dreaded curriculum taught by "technocratic intelligentsia" as a positivistic pursuit of unconnected, specialized knowledge that has contributed to marginalization of other discourses, specifically that of the "humanistic intellectual." Fahlberg and Fahlberg also deconstruct a dreaded curriculum of technical rationality and contrast it with their ideal perspective of emancipatory reason. Similarly, Tinning, Fernández-Balboa, and Brustad describe both dreaded and ideal perspectives within physical education and sport.

I agree with Bain (chapter 11) that, paradoxically, these authors' portrayal of these perspectives in dualistic, dichotomous, and oppositional terms serves to constrain our thinking and limit our ability to find a viable perspective from which to construct a valid position of action. By adopting such an innately modernistic approach, these authors clearly demonstrate how easily it is for the oppressed to become the oppressor. In the following section, I will present the equally dualistic, dichotomous, and oppositional perspective advocated by my technical-rational colleagues to frame an argument against the dreaded emancipatory curriculum in physical education and professional teacher preparation.

The Dreaded Emancipatory Curriculum

For technical-rational disciplinary scholars and scientists, the dreaded curriculum is a socially critical, emancipatory approach to teaching and teacher preparation. Technical rationalists are morally offended by the critical-postmodern perspective that attacks exercise and sport forms as violating the essence of physical education and physical activity. They also deny the charge that most sport is patriarchal, heterosexual, and Euro-centric, and that fitness curricula encourage obsessive, compulsive commitments to exercise. To them, the primary moral violation is the denial of the opportunity to learn and engage in skillful movement and participate regularly in physical activity.

Further, from a practical standpoint, technical rationalists argue that there is simply not enough time in most educational, corporate, community, or commercial programs for students or clients to learn skills and improve performance and, at the same time, critically reflect on the implications of skill and performance enhancement. When a socially critical approach is attached to the performance-focused curriculum, the time that can be allotted for skill development (Barrett 1995) and physical activity to maintain target heart rate is severely limited (McKenzie and Sallis, in press). Therefore, the critical-emancipatory curriculum is particularly dreaded by technical rationalists because the opportunity to learn to perform physical skills and activities proficiently is constrained by a focus on *thinking* about the personal and social meanings and implications of those performances. This issue becomes particularly relevant when time constraints are imposed on the curriculum. For example, the Maryland State Department of Education has reduced the high school physical education requirement to one half credit or one semester of physical education. This means that many Maryland high school students will have their last formal experience with physical education at age fourteen. What should those final forty-five hours of instruction include?

Advocates of a performance-based technical rational curriculum would argue that this curriculum should focus on the development of students' ability to perform movement skills and to acquire the psychological, physiological, and social benefits that advocates believe are minimally necessary to participate in movement (Kirk, chapter 4; Tinning, chapter 7). Currently, this is not limited to simply memorizing facts or performing rote movements, but also includes achieving skillful performance. True, there is often a lack of connection between the biophysical and the sociocultural knowledge both in the nature of the content to be mastered and the definition of mastery. Many biophysical advocates argue that knowledge can be applied to problems regarding the efficiency and effectiveness of human movement, while sociocultural scholars encourage reflection for awareness and insight into the implications and consequences of decisions within a personal and social construction of meaning.

Hence, a technical-rational perspective dreads the emancipatory, socially critical perspective that appears to be void of efforts at serious skill development and strategies necessary to move effectively and participate in an active life-style. Technical rationalists argue that without movement skills, it may be impossible for many individuals to find freedom or transcendence. They also contend that in the dreaded emancipatory curriculum, individuals are deskilled and unable to manipulate objects, move within spatial and temporal constraints, or perform a variety of movements necessary to participate successfully in a variety of movement forms. Specifically, they argue that individuals lack skills to accomplish movement tasks. Despite the ability to think reflectively and critically about sport and fitness-enhancing activities, individuals may not have the knowledge and skills necessary to perform specific physical activities safely and successfully (Clark 1995). Conversely, some postmodern scholars counter that, when individuals are liberated from the necessity of skill development, they can then focus on tactical approaches to sport (Griffin 1995), create opportunities for agency (Hellison and Templin 1991), and achieve true freedom (Fahlberg and Fahlberg, chapter 5).

Technical rationalists, dread the emancipatory curriculum because they perceive that it devalues movement and exercise and deconstructs the role of sport and the value of physical activity in an active, healthy life-style. They argue that the problem is *not* with individuals who abuse physical activity, but with the much larger number of individuals who do not participate in any form of movement and presumably do not find satisfaction with or reason to engage in a physically active life-style (Blair 1993).

When applied to teacher preparation, the emancipatory curriculum is dreaded because it undermines the focus on skillful performance and the enhancement of the quality of movement and thus has the power to negatively affect thousands of students enrolled in physical education (Barrett 1995). As such, technical rationalists fear that preservice teachers no longer will be required to contribute to student learning associated with skillful movement, sport, and exercise, but, instead, will be encouraged to lead extensive discussions regarding participant roles in sport and exercise. They fear that the focus of instruction will be on answering questions such as, "Why should I exercise?" and not "How do I exercise?" They also fear that students will be taught to avoid movement because of the negative moral implications of moving within a patriarchal, heterosexual, Euro-centric perspective or because it will likely lead to an obsessive need to exercise. Further, when the dreaded emancipatory curriculum is limited to "education *through* the physical," it appears to deny access to individuals who are physically talented, engaged in active life-styles, and who find movement to be both liberating and emancipatory. In addition, technical rationalists argue that even if an emancipatory curriculum would contribute to active engagement in movement, there is as yet little information provided by critical-postmodernists regarding how individuals will learn to think reflectively and critically about movement (O'Sullivan, Siedentop, and Locke 1992). They wonder if this is a natural process that will simply occur, given time and attention within the curriculum.

The ancient debate of education *of* the physical *versus* education *through* the physical (Fernández-Balboa, chapter 8) creates a false dichotomy in which moving is juxtaposed with learning about the opportunities and consequences of movement. This is especially unfortunate because individuals within both the dreaded technical-rational and the dreaded-emancipatory perspectives are denied access to the valuable experiences in which the other positively engages.

Teaching preservice teachers and physical education students to reflect on and critique these issues is a time consuming process that is best accomplished over an extended period of time. For instance, Rovegno (1992) examined the problems of working with a student who chose not to or could not reflect. What is the "pedagogical content knowledge" (Shulman 1986) needed to teach students to reflect and critique? What are the criteria for successful reflection? How are tasks designed, arranged, and presented to enhance reflection and critique? If it is possible to reflect or critique incorrectly, should one then "fail" reflection? Although a few scholars (Fernández-Balboa 1995, see also chapter 8) have suggested topics for inclusion in a sociocultural perspective on teacher preparation, with the notable exception of Hellison's (1995) work, the skills and strategies needed to facilitate reflective and critical thinking in preservice teachers and physical education students appear to be missing from the discourse. Without these particulars, it is not difficult for technical rationalists to dismiss this form of instruction as, at best, a "feel-good" approach, and at worst, as *the* dreaded curriculum.

Currently, there appears to be little ground for compromise between the technical rational and the critical-postmodern perspective. Often, access to resources (e.g., allotted instructional time, faculty expertise, and facilities and equipment) provides the venue for the battles between the advocates of each curricular perspective. Surely, neither curriculum is as damaging as its adversaries suggest. In this section, I have used the same rhetoric to portray the emancipatory curriculum as has been used in this text to slash the technical rational perspective. Both present dualistic, dichotomous, oppositional approaches to discourse in which readers may both accept and reject particular aspects.

It seems critically important that we leave the curriculum bashing, dreaded motif and, instead, embrace a more reasonable approach—one that seeks the essence of our personal and professional commitment to the joys and the rewards of moving. I am not suggesting a compromise that dilutes the power of the message, but rather a more reasonable approach to these dualisms in which movers can find a realization of an authentic moving self. In the following section, I will use the work of Burbules (1995) to frame what I believe is a more integrating and conciliatory perspective.

From Rationality to Reasonableness

One of the most formidable dichotomies developed in this text is between rationality and emancipation. Certainly, the foundations of rationality as a dreaded philosophy have been clearly developed, and the constraints imposed by this form of logic succinctly presented.

Advocates of rationality present it as the "fundamental method for the discovery of truth" (Burbules 1995, p. 83). From this perspective, it is assumed that whatever the problems one confronts, rationality provides a supposedly unique procedure for arriving at a correct solution.

Burbules explains that rationality is "based upon logical deduction, strict rules of evidence, and an avoidance of the distorting tendencies of affect; a method of investigation in which the force of correct answers was thought to be rationally, intrinsically, compelling (that is, 'true')" (p. 83). Furthermore, he explains that the overemphasis on the "epistemic function" of reason overshadows the essence of decisions about what to believe and how to act. These decisions, rather than relying on relativistic assumptions, should encourage us to consider alternate forms of reasoning. Clearly, rationality as developed historically within the modernistic framework, and as presented to generations of scholars, has not provided an effective or an empowering avenue for the discovery of universal truths (if in fact they exist).

Authors within this text have effectively deconstructed the naïve and egotistical perspective that prevents individuals from achieving internally-consistent and contextually-relevant perspectives on personal and social truths. To their credit, these authors have been particularly effective in describing the insidious effects of this paradigm within the study and the depreciation of human movement. They have added to the substantial deconstruction of rationality and have extended these efforts to point out their negative effects on human movement. For example, Fahlberg and Fahlberg (chapter 5) and von der Lippe (chapter 3) painstakingly deconstruct rationality and then use the same process to build an argument in favor of emancipation. This points out the continuing reliance on this form of logic. Unfortunately, in this process, they have created an oppositional perspective that disembodies movers from the essence of their subject matter: the opportunity to explore the joys and limitations of their bodies in the act of moving. I do not find either a rational or an emancipatory perspective to be adequate to meet the needs of either the profession or the individual.

Instead, I believe that Burbules' (1995) conceptualization of *reasonableness* provides a more viable perspective that avoids the destructive dichotomies that frequent critiques of dreaded and ideal orientations. Unlike an extreme focus on the narrow perspective of rationality or emancipation, Burbules argues that one should embrace the search for reasonableness. Reasonableness is a pre-modernistic perspective (it predates the Enlightenment) consisting of neither set of dreaded nor ideal constructs. As such, it might escape the purge of both rationality and the "rage against reason" (Burnstein 1988). Postmodern theorists might wish to consider "reasonableness" as a building block to a more practical approach to program design. Burbules (1995, p. 84–85) proposes that the paradox can be confronted and reconstructed using a different way of thinking:

> This different way of thinking about rationality provides the guidance and structure needed for coherent thought in epistemic, practical and moral matters without proclaiming the existence of transcendental and universal standards that are problematic from a postmodern point of view . . . [It is consistent with] the postmodern view that is rooted in *doubt* rather than *denial*. It asks, skeptically, what follows socially and politically from advocating a formal, universal standard of rationality to which people must be expected to conform; it asks who is silenced, who is intimidated, who is excluded when this and this only defines the standard of credible discourse; it holds in suspense an allegiance to any particular mode of thought, when the entire historical and cultural record urges us in the direction of pluralism and tolerance for diversity in these matters. (pp. 84–85)

Reasonableness entails the capacities of the individual to relate to others within particular contexts of practice; not to the regimented adherence to formal rules or procedures of thought. Instead, individuals can be encouraged to supplement skills necessary for logical thought with a contextual understanding of how these skills can be applied in practice. Reasonableness, then, reflects a multidimensional, pluralistic perspective that defies the existence of a single grand narrative of rationality or emancipation because it is constructed within a uniquely situated context.

Burbules (1995, p. 86) argues that, "A person who is reasonable wants to make sense, wants to be fair to alternative points of view, wants to be careful and prudent in the adoption of important positions in life, is willing to admit when he or she has made a mistake and so on." These behaviors are far more complex than what is typically required by rationality or a technical rational viewpoint. Consistent with a postmodern perspective, we can assess them only within a context that is characterized by thoughtful reflection, discussion, and critique.

The research question changes then from a focus on, "What procedures are likely to lead to the truth?" to "When people have tried to understand the truth within their context or situation, what are the general patterns of investigation that they have identified, and how might these be allowed to evolve over time?" (p. 89). Within these patterns of inquiry, the rules of logic may be expected to play an important, yet a limited part within a socially-defined context. Burbules proposes that the most difficult questions are those that describe and define the nature of social contexts within communication, practice, and judgment.

Burbules proposes a set of virtues associated with reasonableness: objectivity, fallibilism, pragmatism, and judiciousness. These virtues not only are concerned with the nature of reasonableness within personal character, practical contexts, and communicative relations, but are fostered and encouraged by the communities and relations with others that create the very contexts in which we make decisions and initiate actions. Burbules envisions these virtues as "flexible aspects of character" (p. 86) that reflect our integrity and sense of self. They are demonstrated in the choices we make and are shaped by our personalities and our interactions with the persons around us.

Objectivity

Unlike the concept of objectivity defined from a positivistic or measurement perspective (Brustad, chapter 6), an objective reasonableness fosters a capacity for commitment, caring, and feeling regarding the other's position. Striving for reasonable objectivity is the process of considering others' opinions and demonstrating tolerance for others' viewpoints. Objectivity might involve an examination of our own biases and capacities to exercise restraint. It is the ability to realize our limitations to fully understand an alternative position while also attempting to ferret out its merits.

When objectivity is conceptualized as a virtue of reasonableness, individuals are encouraged to articulate a pluralistic perspective through the consideration of multiple perspectives. Reasonable individuals, then, come to have a viewpoint only after thoughtfully and sympathetically considering the merits of several viable alternatives:

> We all know persons who have great intellect but who cannot detach their critical capacities sufficiently to hear and consider alternative points of view. To that degree, they lack what I am calling "objectivity." Similarly, persons who can listen to anything and not react critically to it seem to lack a kind of discernment. At both extremes, these people fail to be reasonable. (Burbules 1995, p. 91)

Objectivity requires that individuals have the opportunity for meaningful interactions and relationships with others. Through these, individuals may develop an understanding of others who are different. They can benefit from extended opportunities to make reasonable decisions and carry out mutually constructed plans of actions.

Fallibilism

Within descriptions of the second "virtue" of reasonableness, fallibilism, Burbules reminds us of the fact that we all have experienced failure, error, frustration, and disappointment. We often learn valuable lessons through our mistakes and the trial and error process that structures much of our quest for knowledge. Fallibilism, then, entails a willingness to take risks and to make mistakes. Conversely, withholding commitments, playing it safe, and avoiding mistakes limit our opportunities to learn, grow, and change.

The virtue of fallibilism requires that we both recognize and admit that we can be wrong. It includes our capacity to hear and respond thoughtfully to criticism, to reflect on why we made an error, and to think about how we can avoid it in the future. This contradicts the more typical adversarial viewpoint found in most university settings in which the academy frequently rewards those who vigorously defend their own positions while effectively attacking those of others.

Fallibilism, however, requires more than simply making a mistake and admitting one's error. It suggests a specific form of learning in which individuals actively reconstruct understanding. This occurs as we encounter different viewpoints and interact with individuals from different cultural and economic backgrounds. Hence, we must seek out contexts that support and encourage difference. Burbules argues that often we are more likely to confront personal perspectives when we are shocked by difference. This reaction requires us to rethink our own perspective and to justify or reconceptualize our beliefs and actions. These situations provide opportunities for learning through objectively examining other perspectives while simultaneously reconsidering our own.

Pragmatism

We are more likely to maintain a pragmatic attitude when confronted frequently with important practical problems that tend to drive the process of intellectual, moral, and political development:

Such an outlook is sensitive to the particulars of given contexts and the variety of human needs and purposes. Most important, pragmatism reflects a tolerance for uncertainty, imperfection, and incompleteness as the existential conditions of human thought and action. Yet it recognizes the need for persistence in confronting such difficulties with intelligence, care, and flexibility. (Burbules 1995, p. 94)

Reasonable persons often lack adequate information to discern the correct answer or choice. The situation may be unpredictable and the outcome uncertain. Nevertheless, Burbules argues that a reasonable person is one who can confront problems "with an open mind, a willingness and a capacity to adapt, and persistence in the face of initial failure or confusion" (p. 95). Reasonable individuals are more likely to be able to maintain a pragmatic attitude in educational contexts in which there is not an over emphasis on winning and in which failure or frustration are accepted as part of the conditions of learning and growth. In these environments, cooperative assistance and constructive suggestions are socially and personally acceptable.

Judiciousness

Burbules suggests that judiciousness is the "capacity for prudence and moderation, even in the exercise of reasonableness, itself" (p. 96). From this, it follows that a reasonable person must have the ability to make prudent judgments. "We may not always act reasonably; we occasionally are confronted with a circumstance when not only is it impossible to be reasonable, it is inappropriate" (p. 96). He or she must be able to determine when to be objective and pragmatic and when to adopt an entirely different approach to a problem. Therefore, these heuristics are most helpful when they *guide* choices instead of dictating them (Ingham, chapter 10). A situation in which an individual is in danger, such as when one is being hurt by another, is usually a time for action, not reason.

The key here is being able to hold competing options in balance, accepting tensions as a condition of reflection. Of course, our ability to exercise reasonable judgment is a matter of degree . . . sometimes we can be more reasonable than other times. Thus, Burbules suggests that the virtues of reasonableness can be learned and developed. The heightened awareness of the quest for reasonableness should increase our "sense of interdependence and kinship with others, and provide a healthy counterbalance to the smug confidence that we can answer every question, solve every problem, or resolve every dilemma on our own" (p. 96).

A Reasonable Curriculum

A reasonable curriculum in physical education teacher education is one in which the participants acknowledge and practice the virtues of objectivity, fallibilism, pragmatism, and judiciousness. Unlike the authors in this book who often distance themselves from the performance aspects of physical activity, sport, and exercise, I prefer to frame a more moderate position in which the virtues of reasonableness become the focus of decision making in the human movement curriculum. From this perspective participants will both learn to perform skillfully and to reflect and critique the social context in which they move. While the virtues will be explicitly taught and modified, they will also be infused throughout the teacher preparation methods courses and school physical education programs.

Teacher Preparation

In teacher preparation, these virtues can be presented as part of course work in foundations classes and infused throughout both the biophysical and sociocultural aspects of the program. For instance, preservice teachers can practice objectivity as they refine their own movement skills and assist others in developing skillful movements. They may take both physical and emotional risks as they attempt new movements and strive to challenge themselves to new levels of skillfulness.

Similarly, the virtue of fallibilism is acknowledged as an important part of the challenge of teaching (Schwager, chapter 9). It occurs concomitantly with objectivity as individuals learn to accept the assistance of others and to offer to help others. Evaluation of performance is an important part of this perspective. Unlike technical-rationalist approaches to evaluation based on a standard, narrowly defined criterion, a broader, more individualized definition of performance is present in a reasonable curriculum. Also, individuals are evaluated only after they have had an opportunity to revise and rework their verbal, written, and motoric performances.

In a reasonable curriculum, the experience of learning to teach occurs in an environment in which the virtue of pragmatism is examined on a daily basis. Moreover, unlike with some current models of teacher education in the United States that withhold opportunities to teach children until just prior to preservice teachers' certification, opportunities to interact with children occur frequently, beginning with the first courses taken in the program. A reasonable teacher preparation curriculum includes field experiences with children throughout the program and in a variety of school contexts. This is essential to provide opportunities for preservice teachers to be objective, consider difference, and to make mistakes and try again. This provides them with multiple opportunities to test and refine content and pedagogical skills, make educational judgments that are meaningful to their students within their particular teaching situation, and understand the pragmatic constraints of educational contexts.

More important than the listing of courses in a teacher preparation program, is the tone of the instruction and the efforts to infuse the meaning of the virtues within a reasonable approach to teacher preparation. Inherent in this effort is a focus to enhance both the movement potential of individuals and their understanding of the personal, political, and social implications for moving.

School Physical Education

Children in physical education should also have the opportunity to experience a reasonable curriculum that enhances their ability to perform in an environment in which movement, reflection, and critique occur in an endless spiral. Students need opportunities to learn to perform skillfully and to reflect on and critique their performance in personal, political, and social ways. They also need to learn to listen objectively to others; consider others' opinions, abilities, and interests; and negotiate mutually acceptable courses of action. Further, they need to experience movement, but as the subject of their own exploration and enjoyment of moving.

Teachers who have had the opportunity to take risks, make mistakes within supportive environments, and try new ideas may be more willing to permit their students to experience the virtue of fallibility. The freedom from perfection may be one of several factors essential for transcendental experiences for individuals of all skill levels. These teachers may not only encourage students to be pragmatic in their decisions and choices, but also may provide them with movement and exercise options within a context of safety and personal well being (Schwager, chapter 9). Also, they may warn their students about overly risky activities and over-dependence on exercise which may be detrimental to their emotional or physical selves.

Through a reasonable curriculum, students and teachers can be actively involved in making joint judgments regarding appropriate ways of moving, safe and productive activities, and fair ways to design and implement games in which all have an equal opportunity to participate. Judgments about how to move; how to engage in positive, healthy movement experiences; and how to structure just and empowering moving experiences are central to teaching students about the virtues of reasonableness in physical education.

Closing Remarks

As I consider the debate between rationalists and postmodernists, I am exceedingly wary of their extreme positions. Rarely in my work examining curriculum in school settings or teaching undergraduates and graduates about curriculum does the extreme position seem very objective, pragmatic, or judicious, nor does it permit fallibility. Usually, solutions to the most difficult problems do not occur when individuals stake out a position that is oppositional and then battle for power and an influential voice. Instead, efforts to reach agreements and to set new courses of action occur most frequently when all those involved attempt objectively to consider others' perspectives, realize their own fallibility, and view the world pragmatically. Only after careful reflection can we make judgments that demonstrate an understanding of the context in which we are working.

While the postmodern scholars in this book present a welcome alternative to the rigid rationalist perspective, they must be particularly careful not to establish an equally extreme position that limits their capacity to be objective, fallible, pragmatic, and judicious. Extreme positions often result in the development of programs that, in turn, assume the characteristics of that same dreaded curriculum.

Instead of alienating the very individuals who most need access to objectivity, fallibilism, pragmatism, and judiciousness, a more reasonable approach seeks lines of communication within a deliberate process of curriculum reform. Efforts to find viable alternatives to the current oppressive rationalistic models of curriculum are occurring in many public schools and teacher preparation programs throughout the world. This often requires slow, painstaking, and deliberate negotiations. Instant gratification rarely occurs in curriculum development. In this regard, patience, persuasion, and persistence are valued processes that lead gradually to substantial curriculum change and new opportunities to create multicultural, multi-narrative, and multi-reality approaches to education. A reasonable approach to postmodern curriculum holds promise to enhance and strengthen human movement experiences for a larger constituency.

Questions for Reflection

Chapter 1
Introduction: The Human Movement Profession— From Modernism to Postmodernism

1. What are the bases of modernism?
2. What are the bases for postmodernism?
3. In what ways could you, *a priori*, answer the questions posed in this chapter referring to the human movement profession in the postmodern era?
4. What fundamental and relational questions emerge from your reflection about this chapter?

Chapter 2
Sociocultural Aspects of Human Movement: The Heritage of Modernism, the Need for a Postmodernism

1. What are the modern forms of economic and political inequality and oppression with regards to class, gender, and race?
2. What are the modern symbolic and cultural meanings concerning (*a*) human movement activities and politics? (*b*) human movement activities and economics? (*c*) human movement activities and social class? (*d*) human movement activities and gender? (*e*) human movement activities and race?
3. What are the postmodern possibilities in terms of human movement and politics, economics, social class, gender, and race?

4. What are some important reasons to analyze human movement in the contexts of politics, economics, social class, gender, and race?
5. What fundamental and relational questions emerge from your reflection about this chapter?

Chapter 3
Gender Discrimination in Norwegian Academia:
A Hidden Male Game or an Inspiration
for Postmodern Feminist Praxis?

1. What do the terms "discourse," "difference," and "deconstruction" mean in postmodernism?
2. What roles does orthodox masculine hegemony play in the oppression of women?
3. What are the parameters of antifeminist sexual harassment?
4. In what ways does antifeminist sexual harassment affect women's academic careers and gender-related curriculum issues?
5. What are some of the useful strategies to struggle against gender discrimination proposed here?
6. What are some important reasons to argue for the inclusion of a feminist agenda in academia?
7. What fundamental and relational questions emerge from your reflection about this chapter?

Chapter 4
Schooling Bodies in New Times:
The Reform of School Physical Education
in High Modernity

1. In what ways have school physical education and sport been participants and outcomes of the processes of constructing meanings about the body in modern times?
2. In what ways has *habitus* influenced codifications of corporeal power?
3. What are the historical processes of physical education in Australia?
4. What are some of the aspects of physical education reform in high modernity proposed here?
5. What are some important reasons to advocate physical education reform in high modernity?
6. What fundamental and relational questions emerge from your reflection about this chapter?

Chapter 5
Health, Freedom, and Human Movement
in the Postmodern Era

1. What are the meanings of "emancipatory interest" according to these authors?
2. What are the oppressive and repressive aspects and factors of modern human movement?
3. What are the limits to freedom of modern human movement?
4. What are the possibilities for freedom in postmodern human movement?
5. What are some important reasons to gain emancipation through postmodern human movement?
6. What fundamental and relational questions emerge from your reflection about this chapter?

Chapter 6
A Critical-Postmodern Perspective on
Knowledge Development in Human Movement

1. In what ways have we traditionally studied human movement?
2. In what contexts have we traditionally studied human movement?
3. What aspects and themes have we traditionally studied in human movement?
4. What alternatives do we have in human movement research in the postmodern era?
5. What are some important reasons to argue for new alternatives in human movement research?
6. What fundamental and relational questions emerge from your reflection about this chapter?

Chapter 7
Performance and Participation Discourses
in Human Movement: Toward a Socially
Critical Physical Education

1. What are the issues and differences regarding the performance and participation discourses in human movement?
2. What are some important reasons to argue for a socially critical physical education?
3. What types of professional missions and objectives have been typical of modern human movement activities?

4. What types of professional missions and objectives should one advocate regarding postmodern human movement activities?
5. What are some important reasons for advocating these?
6. What fundamental and relational questions emerge from your reflection about this chapter?

Chapter 8
Physical Education Teacher Preparation
in the Postmodern Era: Toward a Critical Pedagogy

1. What are the fundamental ties among education, physical education, and the larger society?
2. What are the connections among critical pedagogy, literacy, and human movement?
3. What are some crucial reasons for introducing critical pedagogy in postmodern PETE?
4. What themes could one explore critically in PETE concerning human movement?
5. What implications would the adoption of a critical pedagogical philosophy have for PETE?
6. What fundamental and relational questions emerge from your reflection about this chapter?

Chapter 9
Critical Moral Issues in Teaching Physical Education

1. In what ways can *morality* be defined in the context of education?
2. What moral issues and imperatives should physical education teachers consider?
3. What is the process outlined here for preparing morally responsible physical educators?
4. What new roles should schools and universities adopt regarding morality in postmodern physical education?
5. What are some substantial reasons for doing so?
6. What significant and relational questions emerge from your reflection about this chapter?

Chapter 10
Toward a Department of Physical Cultural Studies
and an End to Tribal Warfare

1. What are our field's academic tribes?

2. What are the professional implications for having these tribes?
3. What are the bases and reasons for the creation of a Department of Physical Cultural Studies (DPCS)?
4. What should DPCS's mission, focus, and structure be?
5. What should the contents and educational processes of DPCS?
6. What fundamental and relational questions emerge from your reflection about this chapter?

Chapter 11
Transformation in the Postmodern Era:
A New Game Plan

1. What are the issues raised by Bain concerning the analyses in part I?
2. What are the bases for Bain's practical analysis?
3. What are the foundations for Bain's personal analysis?
4. In what ways do Bain's analyses strengthen and complement the issues raised in part I?
5. What fundamental and relational questions emerge from your reflection about this chapter?

Chapter 12
A Practical Inquiry into the Critical-Postmodernist
Perspective in Physical Education

1. What are the issues raised by Hellison concerning the analyses in part I?
2. What are Hellison's arguments regarding the inclusion of a practical inquiry into the critical-postmodernist perspective in physical education?
3. In what ways does Hellison's analysis strengthen and complement the issues raised in part I?
4. What fundamental and relational questions emerge from your reflection about this chapter?

Chapter 13
Defining the Dreaded Curriculum:
Tensions Between the Modern and the Postmodern

1. What are the issues raised by Ennis concerning the analyses in part I?

2. What reasons does Ennis have for analyzing the dreaded curriculum?
3. How does Ennis's interpret both the concept of *reasonableness* and its components?
4. What would the implications of a reasonable curriculum be for postmodern school and university physical education?
5. In what ways does Ennis's analysis strengthen and complement the issues raised in part I?
6. What fundamental and relational questions emerge from your reflection about this chapter?

Notes

Chapter 2
Sociocultural Aspects of Human Movement:
The Heritage of Modernism, the Need
for a Postmodernism

1. For other analyses of the problematic relationship between the central features of modernism and postmodernism, see S. Aronowitz, Postmodernism and politics, *Social Text* 18: 94–114; J. Baudrillard, *Selected Writings*, ed. M. Poster (Stanford: Stanford University Press, 1988); S. Lash and J. Urry, *The end of organized capitalism*, (Madison: University of Wisconsin Press, 1987).

2. I realize that there are many other widespread and persistent forms of social inequalities that I do not deal with in this chapter. My omission does not imply that I consider them unimportant, but rather that attempting to address all of them would extend this chapter to an unreasonable length. However, many of them will be analyzed and discussed in other chapters of this book.

3. A discourse is a recurring pattern of language about a phenomenon; it is a portrayal of reality, a world view, that becomes part of the normative understandings that frame and shape how the phenomenon is to be understood. Dominant discourses tacitly and explicitly construct reality by governing what is said and what remains unsaid. Thus, privileged discourses sanction specific human interests and regulate human actions because once a discourse becomes institutionally privileged, others are effectively marginalized.

Chapter 3
Gender Discrimination in Norwegian Academia:
A Hidden Male Game or an Inspiration
for Postmodern Feminist Praxis?

1. Although it is true that males are in majority among the professors of sport and physical education, the gender proportions are not adequately reflected in this committee.

Chapter 4
Schooling Bodies in New Times:
The Reform of School Physical Education
in High Modernity

1. The term "high modernity" is drawn from the work of A. Giddens (1990, 1991), who also uses the terms "late modernity" and "radicalised modernity." This terminology is preferred to the notion of "post modernity" since, as this paper seeks to demonstrate, the continuities between the major institutions of modernity and current new times may be stronger than the discontinuities, certainly stronger than use of the prefix "post" would suggest. At the same time, there can be little doubt that modernity has entered a new, accelerated, phase of development, and the notions of "'high," "late," and "radicalised" modernity are intended, following Giddens, to stress this point.

2. Uses of the term "physical culture" in the nineteenth century stressed a holistic view of human beings in which both mind and body and individual and society were conjoined. The term is used throughout this chapter with a similar purpose of overcoming the binaries surrounding physical activity and the body, and seeks to encompass mind *and* body, biology *and* culture, individual *and* collective within the one frame.

Chapter 5
Health, Freedom, and Human Movement
in the Postmodern Era

1. Here we use "movement" as an umbrella term to include extensions in space that may have some relationship with health.

2. By "praxis" we mean a cycle of action *and* reflection (see Freire 1970). In the era of modernism, praxis was reduced to practice which was then a mere technology of manipulation.

3. For a detailed discussion of "myth" in the way we use the term here, see K. Wilber, *Sex, ecology, and spirituality* (Boston: Shambhala, 1995).

4. In our developmental framework, we recognize a variety of levels of freedom. Although Marx popularized the economic and material level of freedom, and Freud popularized the mental-emotional level, there are many more levels to examine with an emancipatory interest (e.g., Wilber 1983).

5. We are not saying the body and mind were ever separate, but that the view during modernism was that they were, in fact, separate. The bridging of this constructed separation is not, therefore, an integration of the bodymind, but rather it is an integration of our view. For the human movement profession, this will likely be one of the dramatic shifts of postmodernism. With the increasing recognition of the body and the mind as one integrated entity, the term "physical education" will be no longer meaningful. It will be increasingly acknowledge that there can be no education that is only physical or that emphasizes the physical only.

6. Although many maps of consciousness include as many stages in the transpersonal as the personal, we are collapsing all of these into the transpersonal for the sake of brevity. Our comments here relate primarily to initial stages in transpersonal development. See F. Vaughan, Discovering transpersonal identity, *Journal of Humanistic Psychology* 25(3): 13–38; and K. Wilber, Odyssey: *Journal of Humanistic Psychology*, 22(1): 57–90.

7. For the reader who has never had a movement-induced peak experience, our claims may seem merely theoretical, if not outrageous. Nonetheless, there is a tremendous body of literature supporting the existence of these experiences, and an entire psychology (transpersonal psychology) arose to study these experiences, no holds barred (Walsh and Vaughan 1993).

Chapter 7
Performance and Participation Discourses in Human Movement: Toward a Socially Critical Physical Education

1. I draw most of my examples from my own experiences in Australia. Accordingly, they are context-specific and may or may not be relevant to the context of other countries. However, the trends experienced in Australia are also obvious in most other Western countries.

2. At this point, it is useful to recognise that the field of human movement is not alone in its reverence of science and its underpinning ideology of technocratic rationality. D. Schön (1983) provides ample evidence that it is the underpinning epistemology of most professions. However, as Schön points out, the training for many professionals does not equip them to cope

with contemporary professional practice that is increasingly unpredictable, complex, situation-specific, and value-laden.

3. I agree with J. McKay, J. Gore, and D. Kirk (1990) when they claim that the fact that scholars in the field of human movement are now among those who are researching the sociology of the body is a most encouraging trend which can help our field move beyond the orthodox, narrow, scientistic way of understanding the body.

4. D. Kirk (1993) tells us that while docility was originally (in Calvinist times) a positive attribute indicating a learner who was teachable, ready to learn, it has acquired a less positive connotation relating to being pliable and malleable into the wealth-making process of capitalism.

5. No one forces me to rise and go for a jog at 6:30 A.M. I am influenced, often in ways that I don't understand or recognise, by the discourses relating to physical activity as an important part of a "healthy" life-style and to the body as a icon of desirability and attractiveness.

6. Also, as I have argued elsewhere (Tinning 1995), there is an unresolved tension here in expecting the human movement profession to adopt a socially critical perspective on the very forces and vested interests that pay for their professional services.

7. Within this context, different interest groups compete for increasingly scarce resources. On a day-to-day basis, our field lives with the competition and tensions between the performance-oriented professions and the participation-oriented professions. The battles are ideological (Lawson 1991b) and deeply embedded in how we think about our field and its particular missions.

Chapter 8
Physical Education Teacher Preparation
in the Postmodern Era: Toward a Critical Pedagogy

1. Any attempt to classify and characterize education and literacy is, in fact, simplistic. At the risk of committing broad generalizations, and for the sake of illustration, I have used these three versions of education to illustrate tendencies and usual patterns in PETE.

2. An important part the agenda of D. Hellison et al.'s program is concerned with social issues. As such, it can be said that it goes beyond humanistic principles and enters the terrain of critical pedagogy.

3. This is not to say that they do not acknowledge the value of technical and humanistic teaching practices; in fact, they use these practices all the time. The difference resides in the ultimate intentions of the pedagogical act— to create a more equitable society and to preserve the natural environment.

4. What I present here are mere examples, not to be taken literally, of how I implement critical pedagogy in PETE. I acknowledge the nature of my own distorted "reading" of critical pedagogy and, hence, do not assume to speak for all critical pedagogues in our field nor do I pretend to possess the secrets to critical pedagogical practice. Different readings of critical pedagogy will lead to different interpretations of it; thus, educators in different programs may implement different versions of critical pedagogy. Critical pedagogy should be adapted to students' personal and contextual needs, not vice versa.

5. For an outstanding example of critical reflection in PETE, see J. Gore, Pedagogy as text in physical education, in D. Kirk and R. Tinning, eds., *Physical education curriculum and culture* (London: Falmer Press, 1990).

Chapter 9
Critical Moral Issues in Teaching Physical Education

1. Each of these illustrations has come from teaching practices I have actually observed in the schools and demonstrate ways in which physical education teachers practice their moral responsibility (e.g., fostering equity, fairness, and self-reliance).

Chapter 10
Toward a Department of Physical Cultural Studies and an End to Tribal Warfare

1. Among my friends are Ken Conner, Ron Cox, Othello Harris, Thelma Horn, Ron Iannotti, Hal Lawson, and Diana Spillman. Others who have assisted me to reach my position include Eric Aikens, Rob Beamish, Peter Donnelly, Steve Hardy, Robert Hutton, John Loy, Robert Monford, and Bob Weinger. I am grateful to Hal Lawson and especially Steve Hardy for their comments on the first draft. I recognize the fine editorial work of Juan-Miguel Fernández-Balboa who condensed my manuscript without taking anything away for the central ideas.

2. While I shall use implicitly the ideas of J. P. Sartre (1956) and M. Weber (1946, 1978), my simplification of their ideas may result in accusations of distortion from the "purists."

3. Much of my argument here has been outlined in more detail by P. Berger, B. Berger, and H. Kellner in *The Homeless Mind* (New York: Vintage Books, 1973).

4. Because of their numerical superiority and authoritative power, it is the technocratic intelligentsia which deserve the full measure of our decon-

structionist endeavors and "knowledge as ideology" strategies. See J. McKay, J. Gore, and D. Kirk, Beyond the limits of technocratic physical education, *Quest* 42(1): 52–76.

5. In a sense, I asked myself to put to one side or bracket my self-defining "-isms," my self-interested well-being, so as to promote the collective well-becoming. In so doing, I can move from the critical to social telesis (collective, purposive human action with an end-state or goal in mind).

6. In building the curricular scaffold (the ideal-type), I am privileging the conceptual and cross-disciplinary orientation of the humanistic intellectual. I am not privileging, however, the humanistic intellectual's tribal claims to prestige. There are times, however, when the ability to conceptualize is of paramount importance and, given the anomie in our field, this is one of them.

7. This may seem to be a contradiction to my earlier statement that our natural scientists work deductively or from within a theory. However, it is not if we take into account the process of specialization in the research endeavor and if we are prepared to admit that the defining paradigm with its more or less middle-range theories has become prereflective in research practices.

8. Thus, there are still many in our field who would argue that the best method is the natural scientific method, with its anchors in explanation, and who look askance at data derived from any other method, especially those with anchors in interpretation.

9. This is the double hermeneutic that the natural scientific method tries to preclude, but must include as research is transformed into policy.

10. Moreover, I find it highly paradoxical that our technocratic intelligentsia operate with a faith in incrementalism (which presumably takes time) while seeking funds from grant agencies which want answers as soon as possible. Surely, what they deliver cannot represent the cumulative knowledge of incrementalist research. A contradiction, indeed.

11. I do not wish to deconstruct functionalism here. I merely wish to say that, in historical reality, we can point to many contradictions that neither have been resolved homeostatically nor have involved rational-purposive action in the successful quest to replace existing systems with new ones.

12. "Habitus," according to P. Bourdieu (1977), is a set of acquired patterns of thought, behavior, and taste that constitutes the link between social structures and social practices.

13. The former refers to the ways in which science and technology are used to standardize the body and movement according to the knowledge gained from the elite performer or ideal type. The latter refers to the ego-defenses we employ when our bodies or our movements "catch us out" (see

Fahlberg and Fahlberg, chapter 5, for a more detailed explanation of the process of ego-defense).

14. I have borrowed much of what follows from my colleagues, especially Sandra Woy-Hazleton, in the Institute of Environmental Sciences at Miami University.

Chapter 11
Transformation in the Postmodern Era:
A New Game Plan

1. In May 1995, the (Wilmington) *News Journal* published a report attributing remarks to CBS golf announcer Ben Wright in which he was alleged to have said that the lesbian presence in women's golf diminished corporate sponsorship and television coverage of the sport. Mr. Wright denied making the remarks and LPGA Commissioner Charles Mechem issued a statement saying, "I do not believe that lesbianism is a significant issue for the LPGA or that is has impeded its growth in the past, or that it will impede its growth in the future." Several players and spokespersons for corporate sponsors also dismissed the significance of Wright's remarks. However, consultants on corporate crisis management noted that sponsors had to downplay the controversy to avoid creating more problems (*USA Today*, May 12, 1995). It should be noted that no one associated with the LPGA said that the presence of lesbians in golf is irrelevant. P. S. Griffin (1992, p. 261) argues that women in sport need to "challenge the use of the lesbian label to intimidate all women in sport."

Chapter 12
A Practical Inquiry into the Critical-Postmodernist
Perspective in Physical Education

1. I use the term critical-postmodernist perspective to encompass the general view espoused by all of these chapter authors. I recognize that there are some differences between critical and post-modernist theory, but since these chapters tend to view the transformation to post-modernism as the foundation for a critical perspective in physical education, I have treated their view as one ideology.

2. I may be misreading R. Tinning, but I believe that physical education can make a contribution to helping people to live more productive and decent lives outside the physical activity setting. However, it is no panacea and does not address, for example, root causes of social problems such as poverty.

Notes

3. I have no major problem being categorized as a humanist, except that the kids in my current programs do make decisions for the betterment of the group, not just the individual (Hellison 1995). This of course still falls short of "social transformation," but it is a start.

4. This course enrolls both students pursuing teacher certification (PETE) and students interested in youth work as a career, as well as some who do not fit either category. For example, this past semester, a criminal justice major who is going to law school and an NBA top draft choice who plans to work with kids during his professional career were enrolled. I support H. Lawson and K. Hooper-Briar's (1994) argument for more integration and cross-training among professionals who work with kids as cited in S. Schwager's chapter.

5. N. Cutforth and I (1992) had some success using practical inquiry in a traditional teaching methods course, but in the curriculum course, barriers such as available time and student commitments to commuting, employment, and so forth, have prevented us from requiring students to field test their models.

References

Abraham, S. (1989). Eating disorders: The past, present and future. Dietitians' Association of Australia, *National Conference Proceedings* 8: 34–35.

Adams, H. (1988). *The academic tribes*. Urbana, IL: University of Illinois Press.

Adler, P. A., S. J. Kless, and P. Adler. (1992). Socialization to gender roles: Popularity among elementary school boys and girls. *Sociology of Education* 65: 169–187.

Agger, B. (1991). Critical theory, poststructuralism, postmodernism: Their sociological relevance. *Annual Review of Sociology* 17: 105–131.

———. (1992a). *The discourse of domination*. Evanston, Illinois: Northwestern University Press.

———. (1992b). *Cultural studies as critical theory: New perspectives on social theory*. Cambridge: University Press.

Alexander, K., A. Taggart, and A. Medland. (1993). Sport education in physical education: Try before you buy, *ACHPER National Journal* 40(4): 16–23.

Allen, D. (1985). Nursing research and social control: Alternative models of science that emphasize understanding and emancipation. *Image: The Journal of Nursing Scholarship* 17(2): 58–64.

American College of Sports Medicine. (1991). *Guidelines for exercise testing and prescription*. Philadelphia: Lea and Febiger.

Anderson, W. (1980). *Analysis of teaching physical education*. St. Louis: C. V. Mosby.

Apple, M. W. (1990). *Ideology and curriculum*. New York: Routledge.

Arendt, H. (1961). *Between past and future*. New York: The Viking Press.

Aron, R. (1968). *Progress and desillusion*. New York: Frederick A. Praeger.

Aronowitz, S. (1987/88). Postmodernism and politics. *Social Text* 18: 94–114.

Aronowitz, S., and H. Giroux. (1985). *Education under siege*. London: Routledge and Kegan Paul.

———. (1991). *Postmodern education*. Minneapolis: University of Minnesota.

Arrighi, G. (1991, September/October). World income inequalities and the future of socialism. *New Left Review* 189: 39–65.

Australian Sports Commission. (1991). *Sport for young Australians: Widening the gateways to participation*. Camberra, ASC.

Bacharach, S., and E. Lawder. (1980). *Power and politics in organizations*. San Francisco, CA: Jossey Bass.

Bain, L. (1976). Description of the hidden curriculum in secondary physical education. *The Research Quarterly* 47(2): 154–160.

———. (1985a). A naturalistic study of students' responses to an exercise class. *Journal of Teaching in Physical Education* 5: 2–12.

———. (1985b). The hidden curriculum reexamined. *Quest* 37: 145–153.

———. (1988). Beginning the journey: Agenda for 2001. *Quest* 40: 96–106.

———. (1989). Interpretive and critical research in sport and physical education. *Research Quarterly for Exercise and Sport* 60(1): 21–24.

———. (1990a). Visions and voices. *Quest* 42(1): 2–12.

———. (1990b). A critical analysis of the hidden curriculum in physical education. In D. Kirk and R. Tinning, Eds. *Physical education, curriculum, and culture: Critical issues in the contemporary crisis* (pp. 23–42). London: Falmer Press.

———. (1993). Ethical issues in teaching. *Quest* 45: 69–77.

Bain, L., and A. E. Jewett. (1987). Future research and theory building. *Journal of Teaching in Physical Education* 6: 346–362.

Bain, L., T. Wilson, and E. Chaikind. (1989). Participant perceptions of exercise programs for overweight women. *Research Quarterly for Exercise and Sport* 60: 134–143.

Bannister, R. (1955). *First four minutes*. New York: Dodd Mead.

Barnet, R. J., and J. Cavanagh, (1994). *Global dreams: Imperial corporations and the new world order.* New York: Simon and Schuster.

Barrett, K. R. (1995). *The role of subject matter knowledge in learning to teach.* Paper presented at the National Conference on Teacher Education in Physical Education, Morgantown, WV.

Barrett, M. (1992). Words and things: Materialism and method in contemporary feminist analysis. In M. Barrett and A. Phillips, Eds. *Destabilizing theory in contemporary feminist debates.* (pp. 201–220). Cambridge: Polity Press.

Baudrillard, J. (1987). Modernity. *Canadian Journal of Political and Social Theory* 11(3): 63–72.

———. (1988). *Selected writings.* Edited by M. Poster. Stanford: Stanford University Press.

Bauman, Z. (1987). *Legislators and interpreters: On modernity, postmodernity, and the intellectuals.* Oxford: Polity Press.

———. (1992). *Intimations of postmodernity.* London: Routledge.

———. (1993). *Postmodern ethics.* New York: Blackwell.

Becker, E. (1973). *The denial of death.* New York: The Free Press.

Bell, L., and N. Schniedewind. (1987). Reflective minds/intentional hearts: Joining humanistic education and critical theory for liberating education. *Journal of Education* 109: 55–77.

Bellah, R. (1985). *Habits of the heart.* New York: Harper and Row.

Benhabib, S. (1994). *Autonomi och gemenskap.* Goteborg, Sweden: Daedalos AB.

Bergen, T. J. Jr. (1994). Culture, character, and citizenship. *Journal of Thought* 29(3): 7–16.

Berger, B., and M. Mackenzie. (1980). A case study of a woman jogger: A psychodynamic analysis. *Journal of Sport Behavior* 3: 3–16.

Berger, B., and A. McInman. (1993). Exercise and the quality of life. In R. Singer, M. Murphey, and L. Tennant, Eds. *Handbook on research in sport psychology* (pp. 729–760). New York: Macmillan Publishing Company.

Berger, P. (1963). *Invitation to sociology: A humanistic perspective.* Garden City: Anchor Doubleday.

Berger, P., B. Berger, and H. Kellner. (1973). *The homeless mind: Modernization and consciousness.* New York: Vintage Books.

Best, S., and D. Kellner. (1991). *Postmodern theory: Critical interrogations.* New York: Guilford Press.

Birrell, S., and C. L. Cole. (1994). *Women, sport, and culture.* Champaign, IL: Human Kinetics.

Blair, S. N. (1993). Physical activity, physical fitness, and health. *Research Quarterly for Exercise and Sport* 64: 365–376.

Bohmer, P. (1992, September). Continued stagnation, growing inequality. *Z Magazine*: 59–62.

Bonheim, J. (1992). *The serpent and the wave: A guide to movement meditation.* Berkeley, CA: Celestial Arts.

Bordo, S. (1990). Reading the slender body. In M. Jacobus, E. Fox-Kellner, and S. Shuttleworth, Eds. *Body / politics: Women and the discourse of science* (pp. 83–112). New York: Routledge.

Bordo, S. (1993). *Unbearable weight: Feminism, Western culture and the body.* Cambridge: Cambridge University Press.

Bourdieu, P. (1977). *Outline of a theory of practice.* Cambridge: Cambridge University Press.

———. (1984). *Distinctions: A social critique of the judgment of taste.* Cambridge: Cambridge University Press.

———. (1986). The forms of capital. In J. Richardson, Ed. *Handbook of Theory and Research for the Sociology of Education* (pp. 241–258). New York: Greenwood Press.

———. (1991). *Language and the symbolic power.* Cambridge and Oxford: Polity Press.

Bourdieu, P., and J. C. Passeron. (1977). *Reproduction in education, society and culture.* London: Sage.

Bradbury, M. (1993). The pornography of sport. *The Age* 10: 6.

Braun, D. (1991). *The rich get richer.* Chicago: Nelson-Hall.

Braverman, H. (1974). *Labor and monopoly capital.* New York: Monthly Review Press.

Brodribb, S. (1992). *Nothing matters: A feminist critique of postmodernism.* North Melbourne, Australia: Spinelix Press.

Broekhoff, J. (1972). Physical education and the reification of the human body. *Gymnasion* 9: 4–11.

Brosio, R. A. (1993). Capitalism's emerging world order: The need for theory and brave action by citizen-educators. *Educational Theory* 43(4): 467–482.

Brower, D. (1994). The archdruit himself. In J. White, Ed. *Talking on the water* (pp. 37–55). San Francisco: Sierra Club Books.

Brustad, R. J. (1993). Who will go out and play? Parental and psychological influences on children's attraction to physical activity. *Pediatric Exercise Science* 5: 210–223.

Bryant, C. (1985). *Positivism in social theory and research*. New York: Macmillan.

Burbules, N. C. (1995). Reasonable doubt: Toward a postmodern defense of reason as an educational aim. In W. Kohli, Ed. *Critical conversations in philosophy of education* (pp. 82–102). New York: Routledge.

Burckes-Miller, M., and D. Black. (1988). Eating disorders: A problem in athletics? *Journal of Health Education* 19(1): 22–25.

Burnstein, R. (1988). The rage against reason. In E. McMullin, Ed. *Construction and constraint: The shaping of scientific rationality* (pp. 189–221). South Bend, IN: University of Notre Dame Press.

Burton-Nelson, M. (1994). *The stronger women get, the more men love football: Sexism and the American culture of sports*. New York: Harcourt Brace and Co.

Butler, J. (1994). Contingent foundations: Feminism and the question of "postmodernism." In S. Seidman, Ed. *The postmodern turn* (pp. 153–170). Cambridge: University Press.

Cahn, S. K. (1994). *Coming on strong: Gender and sexuality in Twentieth-century women's sport*. New York: Free Press.

Callan, E. (1994). Beyond sentimental civic education. *American Journal of Education* 102(2): 190–221.

Carnoy, M. (1994). *Faded dreams: The politics and economics of race in America*. New York: Cambridge University Press.

Carnoy, M., and D. Shearer. (1980). *Economic democracy*. New York: M. E. Sharpe.

Carrigan, T., R. Connell, and J. Lee. (1985). Toward a new sociology of masculinity. *Theory and Society* 14(5): 551–604.

Chase, M. A., and G. M. Dummer. (1992). The role of sports as a social status determinant for children. *Research Quarterly for Exercise and Sport* 63: 418–424.

Cherniin, K. (1981). *The obsession: Reflections of the tyranny of slenderness*. New York: Harper Colophon.

Cherrylholmes, C. (1988). *Power and criticism: Poststructural investigations in education*. New York: Teachers College Press.

Chomsky, N. (1989). *Necessary illusions: Thought control in democratic societies*. Bosto: South End Press.

Clark, J. E. (1995). On becoming skillful: Patterns and constraints. *Research Quarterly for Exercise and Sport* 66: 173–183.

Clark, V., S. N. Garner, M. Higonnet, and K. H. Katrak. (1996). *Antifeminism in the academy*. New York: Routledge.

Clatterbaugh, K. (1990). *Contemporary perspectives on masculinity: Men, women, and politics in modern society*. Boulder, CO: Westview Press.

Cohen, D. K. (1988). *Teaching practice: Plus ça change . . .* (Issue paper 88–3). East Lansing, Michigan: State University, National Center for Research on Teacher Education.

Combs, A. (1981). What the future demands of education. *Phi Delta Kappan* 62(5): 369–372.

———. (1989). New assumptions for teacher education. *Foreign Language Annals* 22(2): 129–134.

Connell, R. (1987). *Gender and power*. Stanford: Stanford University Press.

———. (1990). An iron man: The body and some contradictions of hegemonic masculinity. In M. Messner and D. Sabo, Eds. *Sport, men and the gender order* (pp. 83–96). Champaign, IL: Human Kinetics.

Cornell, R. W. (1991). *Gender and power*. Cambridge and Oxford: Polity Press.

Costa, D. M., and S. R. Guthrie. (1994). *Women and sport: Interdisciplinary perspectives*. Champaign, IL: Human Kinetics.

Crawford, R. (1981). *A history of physical education in Victoria and NSW 1872–1939: With particular reference to English precedent*. Ph.D diss., La Trobe University.

———. (1984). Sport for young ladies: The Victorian independent schools, 1875–1925. *Sporting Traditions* 1: 61–82.

Creedon, P. (1994). *Women, media, and sport: Challenging gender values*. Thousand Oaks, CA: Sage Publications.

Crum, B. (1993). Conventional thought and practice in physical education and physical education teacher education: Problems of the profession and prospects for change. *Quest* 45(3): 339–357.

Culler, J. (1982). *On deconstruction: Theory and criticism after Structuralism*. Ithaca: Cornell University Press.

Cutforth, N., and D. Hellison. (1992). Reflections on reflective teaching in a physical education teacher education methods course. *The Physical Educator* 49: 127–135.

Dagnino, E. (1993). An alternative world order and the meaning of democracy. In J. Brecher, J. B. Childs, and J. Cutler, Eds. *Global visions: Beyond the new world order* (pp. 239–245). Boston: South End Press.

Davies, B. (1994). *Poststructuralist theory and classroom practice.* Geelong: Deakin University Press.

Davis, C., and M. Cowles. (1991). Body image and exercise: A study of relationships and comparisons between physically active men and women. *Sex Roles* 25: 33–44.

DeLorenzo, L. C. (1994). *Teaching as moral stewardship: A guide for teacher preparation.* manuscript, Montclair State University.

Dembo, D., and W. Morehouse. (1994). *The underbelly of the U.S. economy: Joblessness and the pauperization of work in America.* New York: Apex Press.

Dening, G. (1993). *Mr. Bligh's bad language: Passion, power and theatre on The Bounty.* Cambridge: Cambridge University Press.

Department of School Education. (1994). *Review of physical and sport education in Victorian government schools.* Melbourne: Victorian Government Printer.

Derrida, J. (1976). *Of grammatology.* Baltimore: Johns Hopkins University Press.

Dewar, A. M. (1986). *The social construction of gender in a physical education programme.* Ph.D. diss., University of British Columbia, Vancouver, British Columbia.

———. (1987). The social construction of gender in physical education. *Women's studies international forum* 10(4): 453–466.

———. (1991). Feminist pedagogy in physical education: Promises, possibilities and pitfalls. *Journal of Physical Education, Recreation, and Dance* 62(6): 68–71, 75–77.

Dewar, A. M., and T. Horn. (1992). A critical analysis of knowledge construction in physical education. In T. Horn, Ed. *Advances in sport psychology* (pp. 13–22). Champaign, IL: Human Kinetics.

Dewey, J. (1934). *Art as experience.* New York: Minton, Balch and Co.

Dippo, D., and S. A. Gelb. (1991). Making the political personal: Problems of privilege and power in post-secondary teaching. *Journal of Education* 173(3): 81–95.

Dishman, R. (1988). *Exercise adherence.* Champaign, IL: Human Kinetics Books.

Dryfoos, J. (1994). *Full-service schools: A revolution in health and social services for children, youth, and families.* San Francisco, Jossey Bass Inc.

Duckworth, E. (1986). Teaching as research. *Harvard Educational Review* 56: 481–495.

Durkheim, E. (1992). *Professional ethics and civic morals*. London: Routledge.

Dykewomon, E. (1983). Traveling fat. In L. Schoenfielder and B. Weiser, Eds. *Shadow on a tightrope: Writings by women on fat oppression* (pp. 144–154). Iowa City: Aunt Lute.

Eccles, J. S., and R. D. Harold. (1991). Gender differences in sport involvement: Applying the Eccles' expectancy-value model. *Journal of Applied Sport Psychology* 3: 7–35.

Eder, D., and S. Parker. (1987). The cultural production and reproduction of gender: The effect of extracurricular activities on peer-group culture. *Sociology of Education* 60: 200–213.

Ennis, C. D. (1994). Urban secondary teachers' value orientations: Delineating curricular goals for social responsibility. *Journal of Teaching in Physical Education* 13: 163–179.

Ennis, C. D., J. Ross, and A. Chen. (1992). The role of value orientations in curricular decision-making: A rationale for teachers' goals and expectations. *Research Quarterly for Exercise and Sport* 63: 38–47.

Epstein, G., J. Graham, and J. Nembhard. (1993). *Creating a new world economy*. Philadelphia: Temple University Press.

Evans, J. (1993). *Equality, education and physical education*. London: Falmer Press.

Evans, J., and B. Davies. (1993). Post-script: physical education post ERA in a postmodern society. In Evans, J., Ed. *Equality, education and physical education*. London: Falmer Press.

Fahlberg, L. A. (1993). *A critical analysis of the ideology of exercise as a health behavior*. Ph.D. diss., University of Northern Colorado.

———. (in press). Freedom and health promotion. *American Journal of Health Promotion*.

Fahlberg, L. L., and L. A. Fahlberg. (1991). Exploring spirituality and consciousness with an expanded science: Beyond the ego with empiricism, phenomenology, and contemplation. *American Journal of Health Promotion* 5(4): 273–281.

———. (1994). A human science for the study of movement: An integration of multiple ways of knowing. *Research Quarterly for Exercise and Sport* 65(2): 100–109.

Fahlberg, L.L., L. A. Fahlberg, and W. Gates. (1992). Exercise and existence: Exercise behavior from an existential-phenomenological perspective. *The Sport Psychologist* 6: 172–191.

Fahlberg, L.L., J. Wolfer, and L. A. Fahlberg. (1992). Personal crisis: Growth or pathology? *American Journal of Health Promotion* 7(1): 45–52.

Falk, R. (1992). *Explorations at the edge of time: The prospects for world order*. Philadelphia: Temple University Press.

Featherstone, M. (1982). The body in consumer culture. *Theory, Culture and Society* 1(2): 18–33.

Fernández-Balboa, J. M. (1993a). Sociocultural characteristics of the hidden curriculum. *Quest* 45: 230–254.

———. (1993b). *Critical pedagogy in PETE*. Paper presented at the International Seminar of the AIESEP, Trois Rivières, Quebec, Canada.

———. (1994). *Critical pedagogy in teacher education: Why and how*. Paper presented at the annual meeting of the American Educational Research Association in New Orleans.

———. (1995). Reclaiming physical education in higher education through critical pedagogy. *Quest* 47(1): 91–114.

Fernández-Balboa, J. M., and J. P. Marshall. (1994). Dialogical pedagogy in teacher education: Toward an education for democracy. *Journal of Teacher Education* 45(3): 172–182.

Fernández-Kelly, M. P. (1983). *For we are sold: I and my people*. Albany: State University of New York Press.

Fine, G. A. (1987). *With the boys: Little league baseball and preadolescent culture*. Chicago: University of Chicago Press.

Finkelstein, J. (1991). *The fashioned self*. Cambridge: Polity.

Fitzclarence, L. (1990). The body as commodity. In D. Rowe and G. Lawrence, Eds. *Sport and leisure: Trends in Australian popular culture* (pp. 96–108). Sydney: Harcourt Brace Jovanovich.

Foon, A. E. (1987). Reconstructing the social psychology of sport: An examination of issues. *Journal of Sport Behavior* 11: 223–230.

Foucault, M. (1977). *Discipline and punish*. London: Allen and Unwin.

———. (1980a). *Power/knowledge: Selected interviews and other writings, 1972–1977*. New York: Pantheon.

———. (1980b). *The history of sexuality, vol. 1: An introduction*. New York: Vintage.

Fox, K. (1991). Physical education and its contribution to health and well-being. In N. Armstrong and A. Sparkes, Eds. *Issues in Physical Education* (pp. 123–38). London: Cassells.

Fraser, N., and L. J. Nicholson. (1990). Social criticism without philosophy: An encounter between feminism and postmodernism. In L. J. Nicholson, Ed. *Feminism/postmodernism* (pp. 19–39). New York: Routledge.

Freeman, R. B., and L. F. Katz. (1994). Rising wage inequality: The United States vs. other advanced countries. In R. B. Freeman, Ed. *Working under different rules* (pp. 29–62). New York: Russell Sage Foundation.

Freire, P. (1970). *Pedagogy of the oppressed*. New York: Seabury Press.

———. (1985). *The politics of education*. South Hadley, MA: Bergin and Garvey.

Fresia, J. (1988). *Toward an American revolution: Exposing the Constitution and other illusions*. Boston: South End Press.

Freud, S. (1961). *Civilization and its discontents*. New York: W.W. Norton.

Fromm, E. (1955). *The sane society*. Greenwich, CT: Fawcett.

———. (1969). *Escape from freedom*. New York: Avon Books.

Fullan, M. (1982). *The meaning of educational change*. New York: Teachers College Press.

Fürst, E. (1988). *Kvinner i Akademia: inntrengere i en mannskultur?* Oslo: NAVF.

Georgiadis, N. (1992). *Practical inquiry in physical education: The case of Hellison's personal and social responsibility model*. Ph.D. diss., University of Illinois at Chicago.

Gergen, K. J. (1987). Toward self as relationship. In K. Yardley and T. Honess, Eds. *Self and identity: Psychosocial perspectives* (pp. 53–64). New York: John Wiley and Sons.

Giddens, A. (1976). *New rules of sociological method*. London: Hutchinson.

———. (1977). *Studies in social and political theory*. London: Hutchinson.

———. (1982). *Profiles and critiques in social theory*. Berkeley: University of California Press.

———. (1990). *The consequences of modernity*. Oxford: Polity.

———. (1991). *Modernity and self-identity: Self and society in the late modern age*. Oxford: Polity Press.

Gill, D. L. (1992). Gender and sport behavior. In T. Horn, Ed. *Advances in sport psychology* (pp. 143–160). Champaign, IL: Human Kinetics.

Giroux, H. A. (1981) *Ideology, culture and the process of schooling*. Philadelphia: Temple University Press.

———. (1983). Theories of reproduction and resistance in the new sociology of education: A critical analysis. *Harvard Educational Review* 53(3): 257–293.

———. (1988a). *Teachers as intellectuals.* South Hadley, MA: Bergin and Garvey.

———. (1988b). Postmodernism and the discourse of educational criticism. *Journal of Education* 170(3): 5–30.

Giroux, H. A., and P. Freire. (1989). Introduction. In D. E. Purpell, *The moral and spiritual crisis in education* (pp. xiii–xviii). New York: Bergin and Garvey.

Glass, J. (1993). *Shattered selves: Multiple personality in a postmodern world.* Ithaca: Cornell University Press.

Glover, S. (1993). National curriculum comments: Creating opportunity or crisis for physical education and physical educators. *The National Journal of the Australian Council for Health, Physical Education and Recreation* 40(2): 19–21.

Goffman, E. (1951). Symbols of class status. *The British Journal of Sociology* 2: 294–304.

———. (1959). *The presentation of self in everyday life.* Garden City, NY: Doubleday Anchor.

———. (1981). *Forms of talk.* Philadelphia: University of Pennsylvania Press.

Goode, W. (1978). *The celebration of heroes.* Berkeley: University of California Press.

Goodlad, J. (1990). *Teachers for our nations schools.* San Francisco: Jossey Bass, Inc.

———. (1992). The moral dimensions of schooling and teacher education. *Journal of Moral Education* 21(2): 87–97.

———. (1994). *Educational renewal: Better teachers, better schools.* San Francisco: Jossey Bass, Inc.

Gordon, D. M., R. Edwards, and M. Reich. (1982). *Segmented work, divided workers.* Cambridge, England: Cambridge University Press.

Gordon, R. (1968). Transference as a fulcrum of analysis. *Journal of Analytical Psychology* 13(20):109–117.

Gore, J. M. (1990). Pedagogy as text in physical education teacher education: Beyond the preferred reading. In D. Kirk and R. Tinning, Eds. *Physical education, curriculum, and culture: Critical issues in the contemporary crisis* (pp. 101–138). London: Falmer Press.

———. (1992). What we can do for you! What can "we" do for "you?" Struggling over empowerment in critical and feminist pedagogy. In C. Luke and J. Gore, Eds. *Feminisms and Critical Pedagogy* (pp. 54–73). New York: Routlege, Chapman and Hall.

Götz, I. L. (1989). Is there a moral skill? *Educational Theory* 39(1): 11–16.

Gould, S. J. (1981). *The mismeasure of man*. New York: Norton.

Gouldner, A. (1970). *The coming crisis in western sociology*. New York: Basic Books.

———. (1979). *The future of the intellectuals and the rise of the new class*. New York: Seabury.

Graham, G., S. Holt-Hale, and M. Parker. (1993). *Children Moving: A Reflective Approach to 'Teaching Physical Education*. Mountain View, CA: Mayfield Publishing Co.

Gramsci, A. (1971). *Selections from the prison notebooks*. In Q. Hoare and G. Smith, Eds. (pp. 5–23). New York: International Publishers.

Greene, M. (1978). *Landscapes of learning*. New York: Teachers College, Columbia University.

———. (1991). From thoughtfulness to critique: The teaching connection. *Inquiry* 8(3): 1 and 17–21.

———. (1993). The passions of pluralism: Multiculturalism and the expanding community. *Educational Researcher* 22(1): 13–18.

———. (1995). *Releasing the imagination*. San Francisco: Jossey-Bass.

Griffin, L. (1995). *Approach to teaching games and sports: A complete pedagogical content knowledge package*. Paper presented at the National Conference on Teacher Education in Physical Education, Morgantown, WV.

Griffin, P. (1985). Boys' participation styles in a middle school physical education team sports unit. *Journal of Teaching in Physical Education* 4: 100–110.

———. (1992). Challenging the game: Homophobia, sexism, and lesbians in sport. *Quest* 44(2): 251–265.

Griffin, P. S. (1993) Homophobia in women's sports: The fear that divides us. In G. L. Cohen, Ed. *Women in sport: Issues and controversies* (193–203). Newbury Park: Sage.

———. (1984). Girls' participation patterns in a middle school team sports unit. *Journal of Teaching in Physical Education* 4: 30–38.

———. (1985). Boys' participation styles in a middle school physical education sport unit. *Journal of Teaching in Physical Education* 4: 100–110.

Griffith, D. (1993). *Jones's Minimal: Low-Wage Labor in the United States*. Albany, NY: SUNY Press.

Grof, S. (1988). Modern consciousness research and human survival. In S. Grof and M. Valier, Eds. *Human Survival and consciousness evolution* (pp. 57–79). Albany, NY: SUNY Press.

Gruba-McCallister, F. (1991). Behaviorism and existentialism revisited: Further reflections. *Journal of Humanistic Psychology* 31(1): 75–85.

Gruneau, R. (1982). Sport and the debate of the state. In H. Cantelon and R. Gruneau, Eds. *Sport, culture and the modern state* (pp. 1–38). Toronto: University of Toronto Press.

———. (1988). Modernization of hegemony: Two views on sport and social development. In F. Harvey and H. Cantelon, Eds. *Not just a game* (pp. 9–32). Ottawa: University of Ottawa Press.

Haavind, H. (1986). *Kvinneforskning og vitenskapelige paradigmer*. Sekretariat for kninneforskning, no. 1. Oslo: NAVF.

Haber, H. F. (1994). *Beyond postmodern politics*. New York: Routledge.

Hallinan, C. (1991). Aborigines and positional segregation in the Australian Rugby League. *International Review for the Sociology of Sport* 26: 69–81.

Hamilton, L. G. (1941, July). Games practice: Its place and value in the school, *Victorian Education Gazette and Teachers' Aid*.

Hannah, B. (1991). Learning active imagination. In C. Zweig and J. Abrams, Eds. *Meeting the shadow* (pp. 295–297). New York: Tarcher.

Harding, S. (1987). *Feminism and methodology*. Bloomington: Indiana University Press.

———. (1990). Feminism, science, and the anti-enlightenment critiques. In L. J. Nicholson, Ed. *Feminism/Postmodernism* (pp. 83–106). New York: Routledge, Chapman and Hall.

Hargreaves, J. (1986). *Sport, power and culture*. New York: St. Martin's Press.

Harpham, G. G. (1994). So . . . What is Enlightenment? An inquisition into modernity, *Critical Inquiry* 20(3): 524–556.

Harre, R. (1981). The positivist-empiricist approach and its alternative. In P. Reason and J. Rowan, Eds. *Human inquiry: A sourcebook of new paradigm research* (pp. 3–17). New York: Wiley and Sons.

Harris, J. C. (1983a). Broadening horizons: Interpretive cultural research, hermeneutics, and scholarly inquiry in physical education. *Quest* 35: 82–96.

Harris, J. C. (1983b). Interpreting youth baseball: Players' understandings of attention, winning, and playing the game. *Research Quarterly for Exercise and Sport* 54: 330–339.

Hartsock, N. (1990). Foucault on power: A theory for women? In L. J. Nicholson, Ed. *Feminism / Postmodernism* (pp. 157–175). New York: Routledge, Chapman and Hall.

Harvey, D. (1989). *The condition of postmodernity: An inquiry into the origins of cultural change.* Oxford: Basil Blackwell.

Harvey, J., and R. Sparks. (1991). The politics of the body in the context of modernity. *Quest* 43: 164–189.

Hause, S. C. (1984). *Women's suffrage and social politics in the French Third Republic.* Princeton, NJ: Princeton University Press.

Held, D. (1980). *An introduction to critical theory.* Berkeley: University of California Press.

Hellison, D. (1983). It only takes one case to prove a possibility . . . and beyond. In T. Templin and J. Olson, Eds. *Teaching in physical education* (pp. 102–106). Champaign: Human Kinetics.

———. (1985). *Goals and strategies for teaching physical education.* Champaign, IL: Human Kinetics.

———. (1988). Our constructed reality: Some contributions of an alternative perspective to physical education pedagogy. *Quest* 40: 84–90.

———. (1992). If Sargent could see us now: Values and program survival in higher education. *Quest* 44(3): 398–411.

———. (1995). *Teaching responsibility through physical activity.* Champaign, IL: Human Kinetics.

Hellison, D., and T. J. Templin. (1991). *A reflective approach to teaching physical education.* Champaign, IL: Human Kinetics.

Herbert, U. (1990). *A history of foreign labor in Germany, 1880–1980.* Ann Arbor, MI: University of Michigan Press.

Hochschild, A. R., and A. Machung. (1989). *The second shift: Inside the two-job marriage.* New York: Avon.

Hoffer, E. (1951). *The true believer.* New York: Harper and Row.

Hoffman, S. J. (1987). Dreaming the impossible dream: the decline and fall of physical education. In J. Massengale, Ed. *Trends towards the future in physical education* (pp. 121–137). Champaign, IL: Human Kinetics.

Holter, H. (1976). Om kvinneundrtrykkelse, mannsundertrykkelso of hersketeknikker. In T. Støren and T. S. Wetlesen, Eds. *Kvinnekunnskap* (pp. 61–83). Oslo: Gyldendal Forlag.

———. (1991). *Den vitenskapelige bodømmelse: Et knutepunkt for makt of følelser.* In A. Taksdal, Ed. *Veiet og fonnet for lett—og for tung* (pp. 7–19). Arbeidsnotat 1/19. Oslo: NAVFs sekretariat for kvinneforskning.

hooks, b. (1989). *Talking back: Thinking feminist, thinking back*. Boston: South End Press.

hooks, b. (1990). *Yearning: Race, gender, and cultural politics*. Boston: South End Press.

Hooper-Briar, K., and H. Lawson. (1994). *Serving children, youth and families through interprofessional collaboration and service integration: A framework for action*. Oxford, OH, The Danforth Foundation and The Institute for Educational Renewal at Miami University.

Horney, K. (1942). *Self-analysis*. New York: W. W. Norton and Co.

Hoshmand, L. L. S. T. (1989). Alternative research paradigms: A review and teaching proposal. *The Counseling Psychologist* 17: 3–79.

Howze, E., M. Smith, and D. DiGilio. (1989). Factors affecting the adoption of exercise behavior among sedentary older adults. *Health Education Research* 4(2): 173–180.

Ingham, A. G. (1986). *Against knowledge hierarchies and disciplinary segregation*. National Symposium on Graduate Study in Physical Education, National Association for Sport and Physical Education, Cincinnati, Ohio.

Ingham, A. G. and S. Hardy. (1984). Sport, structuration and hegemony. *Theory, Culture and Society* 2: 85–103.

Ingram, D., and J. Simon-Ingram. (1992). *Critical theory*. New York: Paragon House.

Jaggar, A. M. (1983). *Feminist politics and human nature*. Totowa, NJ: Rowman and Allanheld.

Jameson, F. (1988). Cognitive mapping. In C. Nelson and L. Grossberg, Eds. *Marxism and the interpretation of culture* (pp. 347–357). Chicago: University of Illinois Press.

Jameson, F. (1990a). *Postmodernism, or the logic of late capitalism*. Durham, NC: Duke University Press.

———. (1990b). A conversation with S. Hall: Clinging to the wreckage. *Marxism Today* 34: 28–31.

Jay, M. (1984). *Marxism and totality: The adventures of a concept from Lukàcs to Habermas*. Berkeley, CA: University of California Press

Jewett, A., L. Bain, and C. Ennis. (1995). *The curriculum process in physical education*. Dubuque, Iowa: Brown and Benchmark Publishers.

Johnson, D. (1994). *Body, spirit and democracy*. Berkeley: North Atlantic Books.

Jung, C. G. (1971). Aion: Researches into the phenomenology of the self, collected works, Vol. 9. In J. Campbell, Ed. *The Portable Jung* (pp. 139–162). New York: Viking.

Kagan, D., and R. Squires. (1985). Addictive aspects of physical exercise. *Journal of Sports Medicine* 25: 227–235.

Kamel, R. (1990). *The global factory: Analysis and action for a new economic era*. Philadelphia: American Friends Service.

Katz, J. (1986). Long distance running, anorexia nervosa, and bulimia: A report of two cases. *Comprehensive Psychiatry* 27(1): 74–78.

Kemmis, S., and R. McTaggart. (1988). *The action research planner*. Geelong, Australia: Deakin University Press.

Khmeikov, V. T., T. A. Makogon, and F. C. Power. (1995). *The development of self-worth in adolescence: A moral perspective*. Paper presented at the annual meeting of the American Educational Research Association, San Francisco, CA.

Kierkegaard, S. (1849). *The sickness unto death*. London: Penguin Books.

Kirk, D. (1986a). Beyond the limits of theoretical discourse in teacher education: Toward a critical pedagogy. *Teaching and Teacher Education* 2(2): 155–167.

———. (1986b). A critical pedagogy for teacher education: Toward an inquiry-oriented approach. *Journal of Teaching in Physical Education* 5(4): 230–243.

———. (1990). Knowledge, science and the rise and rise of human movement studies. *ACHPER National Journal* 127: 25–27.

———. (1992a). Physical education, discourse, and ideology: Bringing the hidden curriculum into view. *Quest* 44: 35–56.

———. (1992b). *Articulations and silences in socially critical research on physical education: Toward a new agenda*. Paper presented at the AARE Annual Conference, Geelong, Australia.

———. (1992c). Curriculum history in physical education: A source of struggle and a force for change. In A. C. Sparkes, Ed. *Research in physical education and sport: Exploring alternative visions* (pp. 211–230). London: Falmer Press.

———. (1993). Curriculum work in physical education: Beyond the objectives approach? *Journal of Teaching in Physical Education* 12: 244–265.

———. (1993). *The body, schooling and culture*. Geelong: Deakin University Press.

Kirk, D., and B. Spiller. (1993). Schooling for docility-utility: Drill, gymnastics and the problem of the body in Victorian Elementary schools. In Meredyth, D and Tyler, D., Eds. *Child and Citizen: Genealogies of schooling and Subjectivity*. Brisbane: Griffith University.

———. (1994). Schooling the docile body: Physical education, schooling and the myth of oppression. *Australian Journal of Education* 38(1): 78–95.

Kirk, D., and R. Tinning. (1994). Embodied self-identity, healthy lifestyles and school physical education. *Sociology of Health and Illness* 16(5): 600–625.

Kirk, D., and K. Twigg. (1993). The militarization of school physical training in Australia: The rise and demise of the Junior Cadet Training Scheme, 1911–1931. *History of Education* 22(4): 391–414.

———. (1994). Regulating Australian bodies: Eugenics, anthropometrics and school medical inspection in Victoria, 1900–1940. *History of Education Review* 23(1): 19–37.

———. (1995). Civilizing Australian bodies: The games ethic and sport in Victorian government schools, 1904–1945. Sporting Traditions. *Journal of the Australian Society for Sports History* 11(2): 3–34.

Kissling, E. (1991). One size does not fit all, or how I learned to stop dieting and love the body. *Quest* 43(2): 135–48.

Kleinman, S. (1978). Dance/movement therapy: Integrative process. *American Journal of Dance Therapy* 2(2): 13.

———. (1986). *Mind and body: East meets West*. Champaign: Human Kinetics.

Kohn, A. (1994). Grading: The issue is not how but why. *Educational Leadership* 52(2): 38–41.

Kolinsky, E. (1989). *Women in West Germany: Life, work, and politics*. New York: St. Martin's Press.

Kolodny, A. (1996a). Paying the price of antifeminist intellectual harassment. In V. Clark, S. Nelson Garner, M. Higonnet, and K.H. Katrak, Eds. *Antifeminism in the academy* (pp. 3–34). New York: Routledge.

———. (1996b). Why feminists need tenure. *The Women's Review of Books* 13(5): 23–24.

Koplan, J., D. Siscovick, and G. Goldbaum. (1985). The risks of exercise: A public health view of injuries and hazards. *Public Health Reports* 100: 189–195.

Koval, R. (1986). *Eating your heart out*. New York: Penguin.

Krane, V. (1994). A feminist perspective on contemporary sport psychology research. *The Sport Psychologist* 8: 393–410.

Kuhn, T. (1962). *The structure of scientific revolutions*. Chicago: University of Chicago Press.

Kurth-Schai, R. (1992). Ecology and equity: Toward the rational reenchantment of schools and society. *Educational Theory* 42(2): 147–163.

LaBotz, D. (1992). *Mask of democracy: Labor suppression in Mexico today*. Boston: South End Press.

Laclau, E. (1988). Politics and limits of modernity. In A. Ross, Ed. *Universal abandon?* (pp. 63–82). Minneapolis: University of Minnesota Press.

Lasch, C. (1991). *The culture of narcissism*. New York: Norton.

Lash, S., and J. Urry. (1987). *The end of organized capitalism*. Madison: University of Wisconsin Press.

Lavoie, M. (1989). Stacking, performance differentials, and salary discrimination in professional ice hockey. *Sociology of Sport Journal* 6: 17–35.

Lawson, H. (1984). Problem-setting for physical education and sport. *Quest* 36: 48–60.

———. (1988). Occupational socialization, cultural studies, and the physical education curriculum. *Journal of Teaching in Physical Education* 7: 265–288.

———. (1991a). Three perspectives on induction and a normative order for physical education. *Quest* 43(1): 20–36.

———. (1991b). *Specialization and fragmentation among faculty as endemic features of academic life*. Paper presented at the AIESEP World Congress, Atlanta, Georgia.

———. (1993a). After the regulated life. *Quest* 45: 523–545.

———. (1993b). Dominant discourses, problem setting, and teacher education pedagogies: A critique. *Journal of Teaching in Physical Education* 12(2): 149–160.

———. (1994). *International changes and challenges: their import for new models for practice*. Paper presented at the AIESEP World Congress, Berlin.

Lawson, H., and K. Hooper-Briar. (1994). *Expanding partnerships: Involving colleges and universities in interprofessional collaboration and service integration*. Oxford, OH: The Danforth Foundation and The Institute for Educational Renewal at Miami University.

Lebowitz, M. A. (1988). Social justice against capitalism. *Monthly Review* 40(1): 28–37.

Levine, J. M., L. B. Resnick, and E. T. Higgins. (1993). Social foundations of cognition. *Annual Review of Psychology* 44: 585–612.

Lippe von der, G. (1982). Likestilling i idretten. In G. von der Lippe, Ed. *Kvinner og idrett* (pp. 13–37). Oslo: Gyldendal Forlag.

———. (1993). *Heresy as a victorious practice in the field of sport and sexuality in Norway.* Paper presented at the NASS Conference, Canada.

Liston, D., and K. Zeichner. (1991). *Teacher education and the conditions of schooling.* New York. Routledge.

Locke, L. F. (1989). Qualitative research as a form of scientific inquiry in sport and physical education. *Research Quarterly for Exercise and Sport* 60: 1–20.

Lorber, J. (1993). Why women physicians will never be free equals in the American profession. In E. Riska and K. Wegar, Eds. *Gender, work, and medicine: Women and the medical division of labour,* (pp. 62–73). London: Sage.

Loy, D. (1992). Avoiding the void: The lack of self in psychotherapy and Buddhism. *Journal of Transpersonal Psychology* 24(2): 151–180.

Lukàcs, G. (1923/1971). *History and class consciousness.* Translated by R. Livingstone. Cambridge: MIT.

Lyon, D. (1994). *Postmodernity.* Minneapolis: University of Minnesota Press.

Lyotard, J. F. (1984). *The postmodern condition: A report on knowledge.* Minneapolis: University of Minnesota Press.

———. (1992). *The postmodern explained.* Minneapolis: University of Minnesota Press.

———. (1993). *The postmodern explained.* Minneapolis: University of Minnesota Press.

———. (1994). The postmodern condition. In S. Seidman, Ed. *The postmodern turn* (pp. 27–38). Cambridge: University Press.

Macdonald, D. (1992). *Knowledge, power and professional practice in physical education teacher education: A case study.* Ph.D. diss., Deakin University: Geelong.

Maguire, J. (1988). Race and position assignment in English soccer. *Sociology of Sport Journal* 5: 257–269.

———. (1991). Human sciences, sport sciences, and the need to study people "in the round." *Quest* 43: 190–206.

Mangan, J. A. (1975). Athleticism: A case study of the evaluation of an educational ideology. In B. Simon and I. Bradley, Eds. *The Victorian public school* (pp. 147–167). London: Gill and Macmillan.

———. (1986). *The games ethic and imperialism: Aspects of the diffusion of an ideal.* New York: Viking.

Mannheim, K. (1940). *Man and society in an age of reconstruction.* New York: Harcourt, Brace and World.

Marcuse, H. (1970). *Five lectures.* Boston: Beacon Press.

Martens, R. (1987). Science, knowledge, and sport psychology. *The Sport Psychologist* 1: 39–55.

Martinek, T. J., and P. G. Schempp. (1988). An introduction to models of collaboration. *Journal of Teaching in Physical Education* 7: 160–164.

Maslen, G. (1995). Drop in tenured jobs alarms Australian academics. *The Chronicle of Higher Education* (September 15): A43.

Maslow, A. (1954). *Motivation and personality.* New York: Harper and Row.

———. (1969). The farther reaches of human nature. *Journal of Transpersonal Psychology* 1(1): 1–9.

Mauss, M. (1973). Techniques of the body. Translated by B. Brewster. *Economy and Society* 2: 70–87.

May, P. (1974). Foreword. In T. Schoop, *Won't you join the dance?* Palo Alto: National Press Books.

May, R. (1983). *The discovery of being.* New York: Norton.

McCarthy, T. (1981). *The critical theory of Jürgen Habermas.* Cambridge, MA: MIT Press.

McDonald, M. (1992). *Put your whole self in.* Ringwood: Penguin Books.

McFate, K. (1991). *Poverty, inequality and the crisis of social policy: Summary of findings.* Washington, D. C.: Joint Center for Political and Economic Studies.

McInman, A., and J. Grove. (1991). Peak moments in sport: A literature review. *Quest* 43: 333–351.

McIntosh, P. C. (1968). *PE in England since 1800.* London: Bell.

McKay, J. (1991). *No pain, no gain? Sport and Australian culture.* Sydney: Prentice Hall.

McKay, J., J. Gore, and D. Kirk. (1990). Beyond the limits of technocratic physical education. *Quest* 42(1): 52–76.

McKay, J., and D. Rowe. (1987). Ideology, the media and Australian sport. *Sociology of Sport Journal* 4(3): 258–73.

McKenzie, T. L., and J. F. Sallis. Physical activity, fitness, and health-related physical education. In S. J. Silverman and C. D. Ennis, Eds. *Student learning in physical education: Applying research to enhance instruction.* Champaign, IL: Human Kinetics, in press.

McNeill, M. (1988). Active women, media representations, and ideology. In J. Harvey and H. Cantelon, Eds. *Not just a game: Essays in Canadian sport sociology* (pp. 195–211). Ottawa, ON: University of Ottawa Press.

Mehan, H. (1992). Understanding inequality in schools: The contribution of interdependence studies. *Sociology of Education* 65: 1–20.

Messner, M. A. (1988). Sports and male domination: The female athlete as contested ideological terrain. *Sociology of Sport Journal* 5: 197–211.

———. (1992). *Power at play: Sports and the problem of masculinity*. Boston: Beacon Press.

Messner, M. A., and D. F. Sabo. (1994). *Sex, violence and power in sports: Rethinking masculinity*. Freedom, CA: Crossing Press.

Miedzian, M. (1991). *Boys will be boys: Breaking the link between masculinity and violence*. New York: Doubleday.

Miller, J. (1995, May/June). Hard time roll on: Growth and well-being on different tracks. *Dollars and Sense*: 8–9, 38–39.

Miller, T. (1990). Sport, media and masculinity. In D. Rowe and G. Laurance, Eds. *Sport and leisure: Trends in Australian popular culture*. Sidney: HBJ.

Mills, C. W. (1959). *The sociological imagination*. New York: Oxford University Press.

Miracle, A. W., and C. R. Rees. (1994). *Lessons of the locker room: The myth of school sports*. Amherst: Prometheus Books.

Morgan, M. (1991). *Mutant message: Downunder*. Lees Summit, Mo: MM Co.

Morrow, R. A., and C. A. Torres. (1994). Education and the reproduction of class, gender, and race: Responding to the postmodern challenge. *Educational Theory* 44: 43–61.

Mosston, M., and S. Ashworth. (1994). *Teaching physical education*. New York: Macmillan Press.

Munrow, A. D. (1955). *Pure and applied gymnastics*. London: Arnold.

Namenwirth, M. (1986). Science through a feminist prism. In R. Beir, Ed. *Feminist approaches to science* (pp. 18–41). New York: Pergamon Press.

National Eating Disorders Information Center (1988). Exercise can be part of an eating disorder. *Bulletin of Toronto General Hospital* 3(3): 1–2.

Nei til EU. (1994). *Solidaritet eller Union*. Oslo: Nei til EU.

Neumann, E. (1969). *Depth psychology and a new ethic*. Boston: Shambhala.

Nixon, H. (1989). Reconsidering obligatory running and anorexia nervosa as gender-related problems of identity and role adjustment. *Journal of Sport and Social Issues* 13(1): 14–24.

Noddings, N. (1992). *The challenge to care in schools.* New York: Teachers College.

O'Hanlon, T. (1980). Interscholastic athletics, 1900–1940: Shaping citizens for unequal roles in the modern industrial state. *Educational Theory* 30: 89–103.

O'Sullivan, M., D. Siedentop, and L. F. Locke. (1992). Toward collegiality: Competing viewpoints among teacher educators. *Quest* 44(2): 266–280.

Ogden, T. (1982). *Projective identification and psychotherapeutic technique.* New York: Jason Aronson.

Orner, M. (1992). Interrupting the calls for student voice in "liberatory" education: A feminist poststructuralist perspective. In C. Luke and J. Gore, Eds. *Feminisms and critical pedagogy* (pp. 74–89). New York: Routledge, Chapman and Hall.

Oyserman, D., and H. Markus. (1990). Possible selves and delinquency. *Journal of Personality and Social Psychology* 59: 112–129.

Palmer, P. (1989). *Domesticity and dirt: House wives and domestic servants in the U.S. 1920–1945.* Philadelphia: Temple University Press.

Parenti, M. (1995). *Democracy for the few.* New York: St. Martin's Press.

Park, R. (1973). Raising the consciousness of sport. *Quest* Monograph 19, 78–82.

Parliament of the Commonwealth of Australia. (1992). *Physical and sport education.* A report by the Senate Standing Committee on Environment, Recreation and the Arts. Canberra: Parliament House.

Patrick, D., and J. Bignall. (1984). Creating the competent self: The case of the Wheelchair runner. In J. Kotarba and A. Fontana, Eds. *The existential self in society* (pp. 205–221). Chicago: The University of Chicago Press.

Pedersen, T. B. (1994). Vil akademia ha flere kvinner? In *Nytt om kvinneforskning*, Sekretariat for kvinneforskning. Norges Forkningsrad. no. 3.

Peele, S. (1985). *The meaning of addiction: Compulsive experience and its interpretation.* Lexington, MA: Lexington Books.

Peiró, C., and. J. Devís, J. (1993). Innovación en educación física y salud: El estudio de un caso en investigación colaborativa. Paper presented at the conference on *Alternative research in physical education.* Málaga, Spain.

Pelletier, K. (1985). *Toward a science of consciousness*. Berkeley: Celestial Arts.

Perelman, M. (1993). *The pathology of the U.S. economy: The costs of a low-wage system*. New York: St. Martin's Press.

Perkins, K., and L. Epstein. (1988). Methodology in exercise adherence research. In R. Dishman, Ed. *Exercise adherence: Its impact on public health* (pp. 399–417). Champaign, IL: Human Kinetics Books.

Peterson, D. (1994). Doing things right. *Z Magazine* 7: 12–15.

Pettit, A. (1992). On bridges, roads and pathways: A personal journey toward "tikkun." In *Joint AARE / NZARE Conference*. Geelong: Deakin University.

Phillips, A. D., and C. Carlisle. (1983). The physical education teaching assessment instrument. *Journal of Teaching in Physical Education* 2(2): 62–76.

Pollock, K. (1988). On the nature of social stress: Production of a modern mythology. *Social Science and Medicine* 26(3): 381–392.

Postman, N. (1985). *Amusing ourselves to death: Public discourse in the age of showbusiness*. London: Heinemann.

Prain, V., and C. Hickey. (1995). Using discourse analysis to change physical education. *Quest* 47: 76–90.

Purpel, D. (1989). *The moral and spiritual crisis in education: A curriculum for justice and compassion in education*. New York: Bergin and Garvey.

Pyke, F. (1993, June). No base, no pinnacle. *The Australian Council of Health, Physical Education and Recreation Victorian Branch Newsletter* 8.

Quantz, R. A., and T. O'Connor. (1988). Writing critical ethnography: Dialogue, multivoicedness, and carnival in cultural texts, *Educational Theory* 38(1): 95–109.

Rachels, J. (1993). *The elements of moral philosophy*. New York: McGraw-Hill, Inc.

Ranson, S. (1990). Towards education for citizenship. *Educational Review* 42(2): 151–166.

Ranson, S., and J. D. Stewart. (1989). Citizenship and government: The challenge for management in the public domain. *Political Studies* 37: 5–24.

Richardson, V. (1994). Conducting research on practice. *Educational Researcher* 23: 5–10.

Rintala, J. (1991). The mind-body revisited. *Quest* 43: 260–279.

Robertson, I. (1988). *The sports drop-out: A time for change?* Belconnen: Australian Sports Commission.

Robinson, D. W. (1990). An attributional analysis of student demoralization in physical education settings. *Quest* 42(1): 27–39.

Rodríguez, P. (1993). The uprooted from the land. In J. Brecher, J. Childs, and J. Cutler, Eds. *Global visions: Beyond the new world order* (pp. 295–298). Boston: South End Press.

Rogers, C. K. (1961). *Freedom to learn.* Columbus, OH: Charles E. Merrill Publishing Co.

Roth, G. (1989). *Maps to ecstasy.* San Rafael, CA: New World Library.

Rothfield, P. (1986). Subjectivity and the language of the body. *Arena* 75: 157–65.

Rovegno, I. (1992). Learning to reflect on teaching: A case study of one preservice physical education teacher. *The Elementary School Journal* 92: 491–510.

Sabo, D., and M. A. Messner. (1993) Whose body is this? Women's sports and sexual politics. In G. L. Cohen, Ed. *Women in sport: Issues and controversies* (15–24). Newbury Park: Sage.

Sadri, A. (1992). *Max Weber's sociology of intellectuals.* New York: Oxford University Press.

Sage, G. H. (1989). A commentary on qualitative research as a form of scientific inquiry in sport. *Research Quarterly for Exercise and Sport* 60: 25–29.

———. (1990). *Power and ideology in American sport: A critical perspective.* Champaign, IL: Human Kinetics.

———. (1992). Beyond enhancing performance in sport: Toward empowerment and transformation. In *Enhancing human performance in sport: New concepts and developments.* American Academy of Physical Education Papers, No. 25, pp. 85–95. Champaign: Human Kinetics.

———. (1993). Sport and physical education and the new World order: Dare we be agents of social change? *Quest* 45(2): 151–164.

Sampson, E. E. (1977). Psychology and the American ideal. *Journal of Personality and Social Psychology* 35: 767–782.

Sansom, D. (1972). Experiencing: A primary aspect of dance therapy. *Proceedings of the Seventh Annual Conference of the American Dance Therapy Association* 111–113.

Sartre, J. P. (1956). *Being and nothingness.* New York: Washington Square Press.

Scheer, R. (1994, April 25). Welfare or work? *The Nation* 258: 545.

Schempp, P. (1993). The nature of knowledge in sport pedagogy. In *World University Games Conference*. Buffalo, New York.

Schmais, C., and E. White. (1968). Introduction to dance therapy. *Proceedings of a Joint Conference by the Research Department of Postgraduate Center for Mental Health, Committee on Research in Dance and the American Dance Therapy Association* 1–6.

Schmidt, F., and A. Friedman. (1986). *Fighting fair: Dr. Martin Luther King for kids*. Miami, Florida: Grace Contrino Adams Peace Education Foundation.

Schneider, J., and D. S. Eitzen. (1986). Racial segregation by professional football positions, 1960–1985. *Sociology and Social Research* 70: 259–262.

Schön, D. A. (1983). *The reflective practitioner: How professionals think in action*. New York: Basic Books.

———. (1987). *Educating the reflective practitioner*. San Francisco: Jossey-Bass.

———. (1991). *The reflective turn: Cases in and on educational practice*. New York: Teachers College Press.

Schubert, W. H. (1986). *Curriculum: Perspective, paradigm, and possibility*. New York: Macmillan.

Schutz, A. (1967). *The problem of social reality, Collected papers I*. The Hage: Martinus Nijhoff.

Schwager, S. (1992). Relay races: Are they appropriate for elementary physical education? *Journal of Physical Education, Recreation and Dance* 63(6): 54–56.

Scott, J. W. (1994). Deconstructing equality-versus-difference: Or the uses of poststructuralist theory for feminism. In S. Seidman, Ed. *The postmodern turn* (pp. 282–298). Cambridge England: Cambridge University Press.

Seidman, S. (1994, ed.). *The postmodern turn*. New York: Cambridge University Press.

Senge, P. M. (1990). *The fifth discipline: The art and practice of the learning organization*. New York: Doubleday.

Shepard, R. (1984). Can we identify those for whom exercise is hazardous? *Sports Medicine* 1: 75–86.

Sherington, G. (1983). Athleticism in the antipodes: The AAGPS of New South Wales. *History of Education Review* 12: 16–28.

260   References

Shields, D. L. L., and B. J. L. Bredemeier. (1995). *Character development in physical activity*. Champaign, IL: Human Kinetics.

Shilling, C. (1991). Educating the body: Physical capital and the production of social inequalities. *Sociology* 25(4): 653–72.

———. (1993a). *The body and social theory*. London: Sage.

———. (1993b). The body, class and social inequities. In J. Evans, Ed. *Equality, education and physical education* (pp. 55–73). London: The Falmer Press.

Shimada, H. (1994). *Japan's "guest workers": Issues and public policies*. Tokyo: University of Tokyo Press.

Shor, I. (1992). *Empowering education: Critical teaching for social change*. Chicago: University of Chicago Press.

Shulman, L. S. (1986). Those who understand: Knowledge growth in teaching. *Educational Researcher* 15(2): 4–14.

Siedentop, D. (1991). *Developing teaching skills in physical education*. Mountain View, CA: Mayfield.

Siegel, E. (1973). Movement therapy as a psychotherapeutic tool. *Journal of the American Psychoanalytic Association* 21(2): 333–343.

Singhe, K. (1988). Transition to a new consciousness. In S. Grof and M. Valier, Eds. *Human survival and consciousness evolution* (pp. 144–150). Albany, NY: SUNY Press.

Slenker, S., J. Price, S. Roberts, and S. Jurs. (1984). Joggers versus non-exercisers: An analysis of knowledge, attitudes and beliefs about jogging. *Research Quarterly for Exercise and Sport* 55(4): 371–378.

Smeal, G., B. Carpenter, and G. Tait. (1994). Ideals and realities: Articulating feminist perspectives in physical education. *Quest* 46(4): 410–424.

Smith, D. E. (1987). Women's perspective as radical critique of sociology. In S. Harding, Ed. *Feminism and methodology* (pp. 84–74). Bloomington: Indiana University Press.

Soja, E. W. (1987, Number 2). The postmodernization of geography: A review. *Annals of the Association of American Geographers* 77: 289–323.

Soltis, J. F. (1986).Teaching professional ethics. *Journal of Teacher Education* 37(3): 2–4.

South, S. J., and G. Spitze. (1994). Housework in marital and nonmarital households. *American Sociological Review* 59: 327–347.

Sparkes, A. C. (1992). *Research in physical education and sport: Exploring alternative visions*. Bristol, PA: The Falmer Press.

Spring, J. (1974). Mass culture and school sports. *History of Education Quarterly* 14: 483–498.

Stack, S. (1994). Democracy and the quest for community: A pragmatic conception. *Journal of Thought* 29(3): 17–26.

Stallings, J. (1995). *School linked comprehensive services for children and families: What we know and what we need to know.* Joint Publication of the U.S. Department of Education Office of Educational Research and Improvement, and the American Educational Research Association.

Stanley, L. S. (1995). Multicultural questions, action research answers. *Quest* 47: 19–33.

Stanley, W. B. (1992). *Curriculum for Utopia.* Albany: State University of New York Press.

Steinberg, R. (1982). *Wages and hours: Labor and reform in twentieth century America.* New Brunswick, NJ: Rutgers University Press.

Swan, P. (1993). Research paper. Deakin University, Australia.

Tappan, M. B., and L. M. Brown. (1989). Stories told and lessons learned: Toward a narrative approach to moral development and moral education. *Harvard Educational Review* 59(2): 182–205.

Tarnas, R. (1991). *The passion of the Western mind.* New York: Ballantine.

Tart, C. (1987). *Waking up.* Boston: New Science Library.

Theberge, N. (1985). Towards a feminist alternative to sport as a male preserve. *Quest* 37: 193–202.

———. (1991). Reflections on the body in the sociology of sport. *Quest* 43(2): 123–134.

Thomas, J. R., and K. E. French. (1985). Gender differences across age in motor performance: A meta-analysis. *Psychological Bulletin* 98: 260–282.

Tierney, W. G. (1993). *Building communities of difference: Higher education in the twenty-first century.* Westport, CT: Bergen and Garvey.

Tiggeman, M., and B. Pennington. (1990). The development of gender differences in body-size dissatisfaction. *Australian Psychologist* 25(3): 306–13.

Timmer, D. A., D. S. Eitzen, and K. D. Talley. (1994). *Paths to homelessness: Extreme poverty and the urban housing crisis.* Boulder, CO: Westview Press

Tinning, R. (1985). Physical education and the cult of slenderness: A critique. *ACHPER National Journal* 107: 10–14.

————. (1987). *Improving teaching in physical education*. Geelong: Deakin University Press.

————. (1991). Teacher education pedagogy: Dominant discourses and the process of problem setting. *Journal of Teaching in Physical Education* 11(1): 1–20.

————. (1992a). Action research as epistemology and practice. In A. C. Sparkes, Ed. *Research in physical education and sport: Exploring alternative visions*. London: Falmer Press.

————. (1992b). Reading action research: Notes on knowledge and human interests. *Quest* 44: 1–14.

————. (1993a). *Physical education and the sciences of physical activity and sport: Symbiotic or adversarial knowledge fields?* Paper presented at the Congreso Mundial de Ciencias de la Actividad Física y Deporte, Granada, Spain, November.

————. (1993b). Teacher education and the development of content knowledge in physical education teaching. In L. Mesa and J. Vez Jeremias, Eds. *Las didácticas específicas en la formación del porfesorado* (I). España: Tórculo Ediciones.

————. (1994). *The sport education movement: Phoenix, bandwagon or hearse?* Paper presented at the ACHPER/ SPARC Conference, Fremantle, Australia.

————. (1995). We have ways of making you think. Or do we? Reflections on "training" in reflective teaching. In C. Paré, Ed. *Better teaching in physical education? Think about it!* Proceedings of the International Seminar on Training of Teachers in Reflective Practice in Physical Education (pp. 21–52). Deparment des Sciences de l'Activité Physique. Université de Quebéc à Trois Rivières, Canada.

Tinning, R., and L. Fitzclarence. (1992). Postmodern youth culture and the crisis in Australian secondary school physical education. *Quest* 44 (3): 287–3043.

Tinning, R., D. Kirk, J. Evans, and S. Glover. (1994). School physical education: A crisis of meaning, *Changing Education: A Journal for Teachers and Administrators* 1(2): 13–15.

Tomaskovic-Devey, D. (1993). *Gender and Racial Inequality at Work*. Ithaca, NY: ILR Press.

Tsutomu, K. (1991). The political economy of golf. *Japan-Asia Quarterly Review* 22(4): 47–54.

Tucker, L., and K. Maxwell. (1992). Effects of weight training on the emotional well-being and body image of females: Predictors of greatest benefit. *American Journal of Health Promotion* 6(5): 338–344.

References 263

Turner, B. S. (1984). *The body and society: Explorations in social theory*. Oxford: Basil Blackwell.

———. (1992). *Regulating bodies: Essays in medical sociology*. London: Routledge.

Vaughan, F. (1985). Discovering transpersonal identity. *Journal of Humanistic Psychology* 25(3): 13–38.

Veatch, R. (1982). Health promotion: Ethical considerations. In R. Taylor, J. Ureada, and J. Denham, Eds. *Health promotion principles and clinical applications* (pp. 393–404). Norwalk, CT: Appleton-Century-Crofts.

Vertinsky, P. (1992). Reclaiming space, revisioning the body: The quest for gender-sensitive physical education. *Quest* 44: 373–396.

Vertinsky, P., and J. Auman. (1988). Elderly women's barriers to exercise, part I: Perceived risks, *Health Values* 12(4): 13–19.

Vygotsky, L. (1978). *Mind in society: The development of higher psychological processes*. (M. Cole and et al., Eds.). Cambridge, MA: Harvard University Press.

Waite, H. (1985). Playing a different game: Toward a counter-sexist strategy in physical education and sport. *Education Links* 25: 23–5.

Walsh, R. (1980). The consciousness disciplines and the behavioral sciences: Questions of comparison and assessment. *American Journal of Psychiatry* 137(6): 663–673.

———. (1983). The Psychologies of East and West: Contrasting views of the human condition and potential. In R. Walsh and D. Shapiro, Eds. *Beyond health and normality* (pp. 39–66). New York: Van Nostrand Reinhold.

Walsh, R., and D. Shapiro. (1983). In search of a healthy person. In R. Walsh and D. Shapiro, Eds. *Beyond health and normality*. (p. 312). New York: van Nostrand Reinhold Company.

Walsh, R., and F. Vaughan. (1993). Minding our world: Service and sustainability. In R. Walsh and F. Vaughan, Eds. *Paths beyond ego* (pp. 227–231). Los Angeles: Tarcher.

Ward, K. (1990). *Women workers and global restructuring*. Ithaca, NY: Cornell University Press.

Washburn, M. (1990). Two patterns of transcendence. *Journal of Humanistic Psychology* 30(3): 84–112.

Watras, J. (1986). Will teaching applied ethics improve schools of education? *Journal of Teacher Education* 37(7): 13–16.

Weber, M. (1946). Marx Weber. In H. Gerth and C. W. Mills, Eds., and Trans., *Essays in sociology* (pp. 267–301). New York: Oxford University Press.

———. (1947). The theory of social and economic organization. In T. Parsons, Ed. (pp. 11–15, 115–118). New York: The Free Press.

———. (1949). *The methodology of the social sciences*. New York: The Free Press.

———. (1978). Economy and society. In G. Roth and C. Wittich, Eds., Vol. 1 (pp. 24–26). Berkeley, CA: University of California Press.

Weedon, C. (1987). *Feminist practice and poststructuralist theory*. New York: Basil Blackwell.

Wells, J. (1990). Schoolyard remedy: Our costliest sports injury. *The Australian* 14: 32.

Wenner, L. A. (1989). The Super Bowl Pregame Show: Cultural fantasies and political subtext. In L.A. Wenner, Ed. *Media, Sports and Society* (pp. 157–179). Newbury Park, CA: Sage.

Whitson, D. (1986). Structure, agency and the sociology of sport debates. *Theory, Culture and Society* 3: 99–107.

Whitson, D., and D. MacIntosh. (1990). The scientization of physical education: Discourses of performance. *Quest* 42(1): 52–76.

Wilber, K. (1975). Psychologia perennis: The spectrum of consciousness. *Journal of Transpersonal Psychology* 7(2): 105–132.

———. (1977). *A spectrum of consciousness*. Wheaton, IL: The Theosophical Publishing House.

———. (1979). *No boundary*. Boston: Shambhala.

———. (1980). *The Atman project*. Wheaton, IL: The Theosophical Publishing House.

———. (1981). *Up from Eden*. Boston: New Science Library.

———. (1982). Odyssey: A personal inquiry into humanistic and transpersonal psychology. *Journal of Humanistic Psychology* 22(1): 57–90.

———. (1983). *A sociable god*. New York: McGraw-Hill.

———. (1993). Paths beyond ego in the coming decades. In R. Walsh and F. Vaughan, Eds. *Paths beyond ego* (pp. 256–266). Los Angeles: Tarcher.

———. (1995). *Sex, ecology, and spirituality*. Boston: Shambhala.

Williams, N. (1992). The physical education hall of shame. *Journal of Physical Education, Recreation and Dance* 63(3): 57–60.

Williams, R. (1977). *Marxism and literature*. New York: Oxford University Press.

Willis, P. (1980). Notes on method. In S. Hall et al., Eds. *Culture, media, language* (pp. 88–95). London: Hutchinson.

Wittine, B. (1989). Basic postulates for a transpersonal psychotherapy. In R. Valle and S. Halling, Eds. *Existential-phenomenological perspectives in psychology* (pp. 269–287). New York: Plenum.

Wolf, D. L. (1992). *Factory daughters: Gender, household dynamics and rural industrialization in Java*. Berkeley, CA: University of California Press.

Wolff, E. N. (1995). *Top heavy: A study of increasing inequality of wealth in America*. New York: The Twentieth Century Fund.

Woods, N., S. Laffrey, M. Duffy, M. Lentz, E. Mitchell, D. Taylor, and K. Cowan. (1988). Being healthy: Women's images. *Advances in Nursing Science* 11(1): 36–46.

Woods, P. (1979). *The divided school*. London: RKP.

World Bank. (1990). *World development report 1990: Poverty*. New York: Oxford University Press (for the World Bank).

Worldwatch Institute. (1990). *State of the world 1990*. New York: W. W. Norton.

Wyss, B., and R. Balakrishnan. (1993). Making connections: Women in the international economy. In G. Epstein, J. Graham, and J. Nembhard, Eds. *Creating a new world economy* (pp. 421–435). Philadelphia: Temple University Press.

Yates, A., K. Leehey, and C. Shisslak. (1983). Running—An analogue of anorexia? *New England Journal of Medicine* 308: 251–255.

Zeigler, E. F. (1986). The influence of ecology on sport and physical education. In E. F. Zeigler, Ed. *Assessing sport and physical education* (pp. 370–391). Champaign, IL: Stipes Publishers.

Archival Materials

Argus, April 1917.

Australian Military Forces. (1916). *Junior cadet training manual*. Melbourne: Government Printer.

Department of Defence. (1922). *Junior cadet training textbook*. Melbourne: Government Printer.

Education Department of Victoria. (1946). *Physical education for Victorian schools*. Melbourne: Government Printer.

Victorian Parliamentary Papers. (1890). *Reports of the Minister of Public Instruction, 1889–90, 95*, xxii.

Victorian Parliamentary Papers. (1929, December). *Reports of the Minister of Public Instruction, 1928–9, 134*.

Victorian Education Gazette and Teachers' Aid. (1933).

About the Authors

Linda L. Bain is Provost and Academic Vice President at San José State University (U.S.A.). She earned her Ph.D. from the University of Wisconsin at Madison. Dr. Bain formerly held administrative positions at California State University, Northridge and the University of Houston. She has held faculty positions at the Universities of Houston, Illinois at Chicago, and Michigan. Her area of specialization is curriculum theory and pedagogy. She has authored numerous books, book chapters, and articles and has been recognized nationally and internationally for her scholarship. She is a Fellow of the American Academy of Kinesiology and Physical Education.

Robert Brustad received his doctorate from the University of Oregon. Currently he is an Associate Professor at the University of Northern Colorado (U.S.A.). He has published in several journals. His specialties are sport psychology and sociology. He is the Associate Editor of the *Journal of Sport and Exercise Psychology*.

Catherine D. Ennis is an Associate Professor in the Department of Kinesiology at the University of Maryland (U.S.A.). She has a Ph.D. from the University of Georgia. She has conducted research examining the influence of teachers' values and beliefs on their curricular decision making. She is a coauthor of *The Curriculum Process in Physical Education* (1995), and is a coeditor of *Enhancing Student Learning in Physical Education*. She has published more than forty research articles and belongs to the editorial boards of several prestigious journals. Dr. Ennis is the recipient of Celebration of Teaching Award, and was the keynote speaker at the 1991 NASPE conference; she has served as Chair of the Curriculum and Instruction Academy (1990).

Larry L. Fahlberg, Ph.D., is an Associate Professor of health education in the School of Physical and Health Education at the University of Wyoming (U.S.A). He is the author of numerous articles and book chapters on freedom and optimal well-being, and serves on several editorial and review boards for health journals. He is currently writing a book entitled *Body, Mind, Spirit: Freedom in Motion*.

Lauri A. Fahlberg, Ed.D., is a clinical adjunct professor in the College of Health Sciences at the University of Wyoming (U.S.A). She has coauthored many articles and book chapters on health and human movement, and has most recently worked as a movement therapist in a psychiatric hospital. In addition to treatment, her current interests include the use of movement for personal and collective growth.

Juan-Miguel Fernández-Balboa, originally from Barcelona (Spain), obtained his Ed.D. in Education from the University of Massachusetts-Amherst (U.S.A). He is an Associate Professor in Physical Education Pedagogy at the University of Northern Colorado (U.S.A.) where he teaches qualitative research, sociology of sport, secondary school physical education curriculum, and critical pedagogy. His interests focus on teacher development, critical theory, and critical pedagogy, themes about which he has written several theoretical and research-based articles in various international journals. He is a member of two editorial boards and a reviewer for several journals.

Don Hellison is a Professor of Kinesiology at the University of Illinois at Chicago (U.S.A). He is most well known for his work with at-risk and urban youth and for the development of affective approaches to teaching physical activities in schools and social agencies. He recently received the "International Olympic Committee's President's Prize" (1995), the University of Illinois at Chicago "Excellence Award" (1995), and the AAHPERD C.D. Henry Award for service to minorities (1994). He has written five books, the most recent *Teaching Responsibility Through Physical Activity* (1995). He currently serves as the editor of *Quest*.

Alan G. Ingham, born in Manchester, England, received his Ph.D. in Sociology at the University of Massachusetts. He is currently a Professor of Sport Studies and Coordinator of Graduate Studies in the Department of Physical Education, Health and Sport Studies at Miami University (U.S.A). Dr. Ingham has served on several editorial boards and on the Executive Board of the International Committee of the Sociology of Sport and was president of this association between 1984 and 1987. He has published many articles and books chapters. He is the coauthor of *Sport in Social Development: Traditions, Transitions, and Transformations*. In addition, he has been the key note and/or plenary speaker at Olympic and Commonwealth Scientific Congresses, Symposia of the International Committee for the Sociology of Sport, and at two of the North American Society for Sport History's annual conferences.

About the Authors 271

David Kirk, a professor of Human Movement Studies at The University of Queensland (Australia), has authored *The Body, Schooling and Culture* (1993), and coauthored *Learning to Teach Physical Education* (1993). He is also a prolific article writer and a member of the editorial boards of several international journals. His current research projects include an evaluation of a new senior school physical education syllabus in Queensland, the emergence of forms of physical culture in Australia since 1945, the emergence of human movement studies in tertiary institutions since the early 1970s, and their influence on social construction of school physical education.

Gerd von der Lippe is an Associate Professor in sports pedagogy, history, and sociology at the Høgskolen i Telemark (Norway). She is the author of two books on women and sport and another on the politics of sport. She has also written several articles in international journals including International Review for the Sociology of Sport, the Journal of Comparative Physical Education. She has written chapters in several books. In addition, Professor von der Lippe is a commentator on sports and politics for two Norwegian newspapers as well as a guest in television programs on women and sport.

George Sage is an internationally renown sport sociologist. He is an Emeritus Professor at the University of Northern Colorado (U.S.A.). Dr. Sage has received numerous awards for his scholarship including selection as the 1985–86 Alliance Scholar for the American Alliance for Health, Physical Education, Recreation and Dance (AAHPERD) and the 1986 National Association for Sport and Physical Education (NASPE) Distinguished Achievement Award. Besides a great number of articles, he has authored *Power and Ideology in American Sport*, and is coauthor of *Sociology of North American Sport*.

Sue Schwager is a Professor of Physical Education at Montclair State University (U.S.A.). She earned her Ph.D. from Teachers College, Columbia University (U.S.A.). Her scholarly interests include the study of teacher development and socialization. Most recently, she has focused her work on critical thinking, critical pedagogy, and the moral implications of teaching and teacher decision-making. She is the author of several articles in these areas.

Richard Tinning holds a personal chair in physical education at Deakin University (Australia). He teaches curriculum and pedagogy courses in preservice, inservice, and post graduate levels. His research interests are in the nature of physical education, knowledge production and appropriation in teacher education, action research, and identity construction and youth culture. He is the author of *Improving Teaching in Physical Education* (1987), and has coauthored *Physical Education, Curriculum & Culture: Critical Issues in the Contemporary Crisis* (1990); *Daily Physical Education: Collected Papers on Health Based Physical Education in Australia* (1991); and *Learning to Teach Physical Education* (1993). He has also published widely in international journals.

Index

273

critical theory, 6, 68, 195
critical thinking, 133, 173
cultural activities, 21
cultural capital, 41
cultural meanings, 21–26
curriculum, 165–168, 174–180, 205,
207–220

D
deconstruction, 27–28
democracies, 15
democracy, 20
difference, 27–28
discipline, 42, 111
discourse-s, 27, 37, 96, 101–102,
106, 122, 165
discriminatory practices, 32, 131,
194
docility, 42, 47, 112
domination, 68
doxa, 28
dreaded curriculum, 207–212, 220,
225–226

E
economic rationalism, 113
economics, 4, 18, 30
economy, 20, 22
education, 121–122, 126, 140, 144,
211
ego identity, 74–75, 78, 83
emancipation, 213
emancipatory curriculum, 209–212
emancipatory interest, 65, 69,
71–72, 86, 223
emancipatory reason, 65–66, 69,
72–73
empirical, 67
empowerment, 173, 199
Enlightenment, 6, 12, 185, 214
equality, 23, 128
equity, 142
ethics, 128
ethnic minorities, 25
existential identity, 75, 81
experimental methodologies, 88

F
fallibilism, 216, 218, 220
female academicians, 32
feminist perspectives, 35, 189
feminist theories, 33
freedom, 65, 68–71, 85–86, 121,
150, 197, 201, 210, 223
functionalism, 168, 172–173

G
games ethic, 48
games, 50
games-playing, 44, 49
gender, 14, 18, 24, 92–93, 123, 221

H
habitus, 40–41, 44, 49, 61–96, 174,
222
health behavior, 66, 129
health, 51, 65–66, 68–70, 73, 82, 86,
95, 114–115, 164, 166, 171, 223
"healthy", 68, 112
healthy bodies, 59
hegemonic epistemologies, 167
hegemonic masculinity, 59–60, 171,
177
hegemony, 22–23, 27–30, 169–170,
173, 175, 186
Hermeneutics, 168
hidden curriculum, 93, 127, 147
high modernity, 40, 42, 56, 61, 63,
111, 222
higher education, 30
human capital model, 106
human development, 77
human movement, 21, 24, 65, 67,
86–87, 99, 102, 108, 115, 197,
210, 223, 224
human movement practices, 23
human movement profession-als, 7,
11, 105–107, 191, 221
human movement programs, 186
human movement research, 88
humanistic educators, 125
humanistic intellectuals, 161–163,
169, 172, 208

I
identity, 74, 177–178, 189, 194
ideology, 22–25, 105, 122, 124, 142, 168, 197
immortality projects, 79
Industrial Revolution, 16
inequality, 13, 14, 15, 17–18, 40, 221
instrumental positivism, 171
instrumental rationality, 158
intellectual harassment, 31

J
judiciousness, 217, 220
justice, 121, 128, 190, 201

K
knowledge development, 87–88
knowledge production, 29
knowledge, 6, 72, 107–108, 116

L
lesbian-s, 193–194
literacy, 124, 224

M
manliness, 49
masculine hegemonic orthodoxy, 27–31, 33–34, 222
masculine, 24
masternarratives, 7, 188
media representations, 130–131
medical assessment, 48
medical inspection, 47–48
metanarratives, 28
methodological practices, 133–137
modern body, 61
modern institutions, 4
modern project, 5
modern society, 11, 25, 222
modernism, 8, 11, 56, 68, 185, 221
modernist discourses, 110
modernistic ideology, 122
modernistic perspective, 214
modernity, 12, 14, 39
moral imperatives, 141–143

moral issues, 139–155, 224
moral physical education, 144
moral responsibilities, 146, 155, 191
moral stewardship, 144, 154
moral-s, 128
morality, 140–141
movement culture, 105
movement practices, 96
movement, 70, 78, 177–178

N
natural environment, 132
neoconservatism, 188
new times, 39, 63
normative order, 147–148
nurturing pedagogy, 151

O
objectivity, 88, 91, 215–216, 220
occupational hierarchy, 159–162
oppression, 13, 16, 20, 69, 71–73, 84–85, 105, 187–188, 201, 221–222
orthodox science, 88, 92
orthodoxy, 27–29, 222

P
paradigm-s, 88–92, 167, 203
participation discourses, 99, 102, 223
participation oriented missions, 114
patriarchal traditions, 4
patriarchy, 14
peak experiences, 83–84
pedagogical practices, 47, 192
pedagogy, 53, 123, 179
performance, 218
performance discourses, 96, 99, 102–103, 110, 164, 223
performance enhancement, 95, 209
performance oriented ideology, 124
performance oriented missions, 113–114
persona identity, 75–76
personal identity, 194
physical activity, 92–93, 211

physical appearance, 95
physical capital, 40
physical cultural studies, 157,
    164–180, 224–225
physical culture, 58–60, 165, 174,
    176
physical education teacher educa-
    tion (PETE), 112, 116, 118, 121,
    123–138, 155, 191, 199–200,
    204–205, 224
physical education, 11, 21–22, 25,
    34, 39, 47–48, 50, 52, 57, 62–63,
    104–107, 139, 144, 152, 185, 199,
    220, 223–225
physical educators, 52, 58, 105, 126,
    147–148, 191
physical training, 44, 46–47
political, 13, 16
popular culture, 21
positivism, 91, 185
positivist model, 89, 91, 164, 208
positivist paradigm, 89–92
positivist science, 88, 97
postmodern era, 5, 13, 65, 69, 115,
    121, 223
postmodern perspective, 192, 208,
    214, 223
postmodern society, 20
postmodern theorists, 5, 214, 220
postmodern thought, 154, 183
postmodernism, 5, 8, 11, 20, 26, 36,
    97–98, 112, 118, 144, 157, 183,
    186, 189, 195
postmodernists, 36, 188
postmodernity, 14, 185
poststructuralism, 118
poststructuralist analysis, 119
poststructuralist theory, 118
power, 27, 29, 42, 79, 123
practical inquiry, 197–205
practices, 163, 168, 171
pragmatic rationalism, 158
pragmatism, 216–218, 220
praxis, 27, 36, 67, 73, 77, 85, 195,
    199, 222
prescriptions, 46, 67, 73, 201

preservice teachers, 211, 219
primary education, 50
primary school, 52
professional knowledge, 110
professional missions, 113–114,
    158, 162–166
professional practice-s, 114, 117,
    164
professional preparation, 198
professional training, 109, 117
projection, 76–77
promotion, 31
psychopathology, 69
public discourse, 104

Q
qualitative paradigm, 90
qualitative research, 90

R
race, 19, 25, 123, 221
racial, 15
racism, 15
radical scholars, 30
rationality, 212–213
reality, 6, 89
reasonable curriculum, 218–220
reasonableness, 212–220, 226
reductionism, 88
reflection, 71, 85, 146, 212
reflective practices, 126, 128–133
regulation, 117
repression, 69, 71, 73, 201
research, 73, 87, 117, 202–204
research paradigms, 88
researchers, 54

S
scholarly interests, 174
school physical education, 39–40,
    53, 56–58, 60, 111, 118, 142, 147,
    190, 218–220
school sport, 25, 40, 49, 56
school-s, 22, 48, 126, 151, 153
science, 67, 110, 116, 167, 185
scientific discourse-s, 33, 116, 132